BEFORE AND AFTER THE I-BOMB:

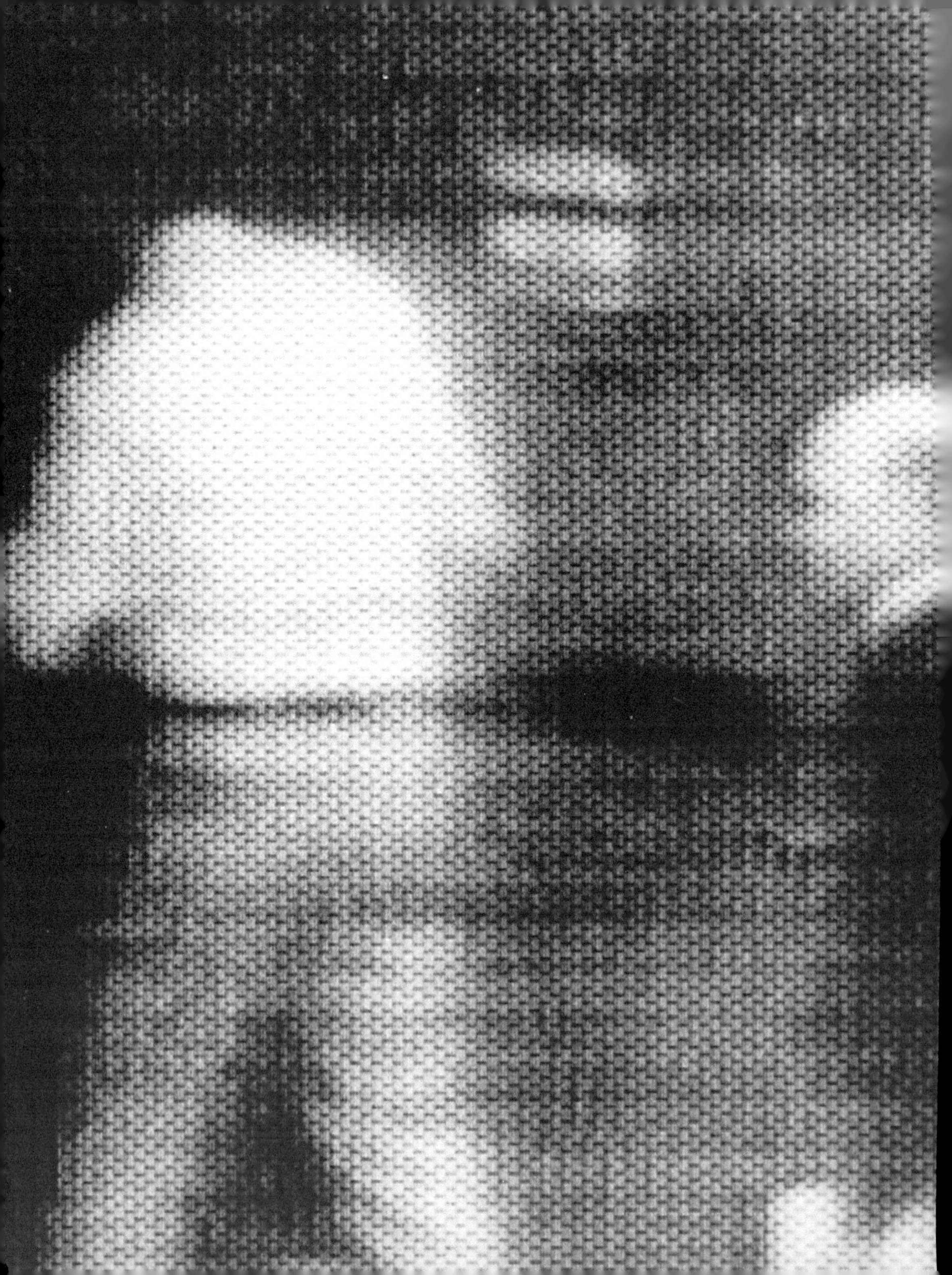

AN ARTIST IN THE INFORMATION ENVIRONMENT
TOM SHERMAN

EDITED BY PEGGY GALE

National Library of Canada Cataloguing in Publication Data

Sherman, Tom.
Before and after the I-bomb

ISBN 0-920159-94-X

I. Gale, Peggy, 1944- II. Title.
PS8587.H3864B43 2002 C814'.6 C2002-910142-5
PR9199.3.S513B43 2002

Edited by Peggy Gale
Cover and book design: Vangool Design + Typography
Printed and bound in Canada by: Houghton Boston Printers

Cover photograph by Tom Sherman
Photo credits: p. 210, courtesy Kurt Hentschläger (Granular Synthesis); p. 212, Chambers reproduction courtesy Art Gallery of Ontario; p. 224, courtesy V Tape, Toronto, Ontario; p. 229, courtesy Les Levine; p. 235, courtesy Susan Britton; p. 246, courtesy Kurt Hentschläger (Granular Synthesis). All other images by Tom Sherman.

The Canada Council for the Arts since 1957 | Le Conseil des Arts du Canada depuis 1957

The Banff Centre Press gratefully acknowledges the Canada Council for the Arts for its support of our publishing program.

The Banff Centre Press

Banff Centre Press
Box 1020 Banff, Alberta T1L 1H5
www.banffcentre.ca/press

> For my parents, Alyce and Ben

TOM SHERMAN

Acknowledgements

I dedicate this book to my parents, Alyce and Ben Sherman. They have always been unconditionally supportive of me and genuinely interested in my work as an artist and writer.

I've been privileged to share perspectives of the world with Jan Pottie, my wife and intellectual companion of twenty-three years. Her intellectual curiosity and breadth of knowledge is inspiring. Together we have raised a daughter, Yolande. As Yolande has grown up and become her own person, she brings her natural gifts to us every day: her ear for music and language, and her wonderful, discerning eye. Jan and Yolande, in their own ways, have helped me write this book.

Thanks to Peggy Gale, my editor, for making this book better. Peggy has been a valued friend and colleague for many years. It has been a pleasure to work with her again, to benefit from her intelligence, insight, and careful attention.

The texts included in this book also bear the influence of many minds, of friends, collaborators, and valued readers. Thanks to Robert Adrian, Roy Ascott, Daina Augaitis, Renée Baert, Kevin Barnhurst, August Black, René Blouin, Joe Bodolai, Jean-Guy Bossé, Susan Britton, AA Bronson, Colin Campbell, James D. Campbell, Spencer Cathy, Guillermo Cifuentes, Lynne Cohen, Robin Collyer, Greg Curnoe, Scott Didlake, Diana Domingues, R. Bruce Elder, Edgar Endress, Jerry Ferguson, Ingrid Fischer-Schreiber, Don Foresta, Peggy Gale, Eldon Garnet, Joy Garnett, Patricia Gruben, Heidi Grundmann, Serge Guilbaut, Susan Harrison, Jocelyn Harvey, Kurt Hentschläger, Willard Holmes, Stephen Horne, Bill Jones, Robert Kennedy, Richard Kerr, Franz Kramer, Richard Kriesche, Denis L'Espérance, Ulf Langheinrich, Jim Leonard, Steven Leuthold, Les Levine, Ardele Lister, Donal Little, Bernhard Loibner, Andrew Lugg, Brian MacNevin, Tanya Mars, Ed Marsjewski, Roger Mayer, Manfred Mixner, Brian Molyneaux, Jason Mombert, Ian Murray, Barry Nevitt, John Orentlicher, John Oswald, Hélène Papineau, Cary Peppermint, Jean Piché, Steve Radlauer, Barak Raefeli, Heimo Ranzenbacher, Michael Recht, Steve Reinke, Clive Robertson,

David Ross, Paul Ryan, Jutta Schmiederer, Christine Schöpf, Tom Shelton, Ron Shuebrook, Ed Slopek, Michael Snow, Diane Spodarek, Jack Stanley, Lisa Steele, Gerfried Stocker, Michael Tarantino, Dennis Tourbin, Elke Town, Biff Ward, Randy Ware, John Watt, Norman White, Michael Williamson, Joe Wright, and Jay Yager, and to the many individuals comprising Nettime, Rhizome Raw, and Thingist listservs, the network-spaces where many of my texts since 1995 were conceived and made public.

Thanks to Timothy Porteous, former director of the Canada Council, who in 1981 made the decision to hire me as video officer of the Council's Visual Arts Section. In 1983, he encouraged and permitted me to found the Media Arts Section, appointing me its first head of section and allotting new resources for computer-integrated media. Through my work at the Council from 1981 to 1987 I learned about the country called Canada.

Thanks to Syracuse University and Don Lantzy, then dean of SU's College of Visual and Performing Arts, and my colleagues at CVPA's School of Art and Design and the Department of Art Media Studies for encouraging and supporting my work throughout the 1990s, and for granting me a study leave that permitted me to organize this book. Thanks also to my students for their wonderful, fresh perspectives on art, media, technology, and life.

And thanks to The Banff Centre for the Arts and the Banff Centre Press for publishing this book. Jon Tupper, then Director of the Walter Phillips Gallery, responded positively to the manuscript and encouraged me to approach the Press. Thanks to Gerald Hushlak, head of the Art Department, University of Calgary, and Sara Diamond, director of Visual and Media Arts and the New Media Institute, The Banff Centre, two particularly enthusiastic readers; Lori Burwash at the Banff Centre Press for keeping the evaluation process going and keeping me in the loop; and Maureen Nicholson, copy editor, for improving my prose. And, finally, thanks to Lauri Seidlitz, managing editor of the Banff Centre Press, for her acceptance of the manuscript for publication.

TOM SHERMAN

A Note to the Reader

PEGGY GALE

These brief words of welcome are perhaps not a full introduction. They offer no translation or mediation of Tom Sherman's writing. Rather, they are an invitation to pause, to step forward and enter the text directly. No external guide is necessary.

In his essay "Explicit Text," Sherman recalls a time when "Reading used to be seeing and thinking. The text object was rigidly fixed, like a drawing, placed in waiting for someone to stumble upon it, to pore over it, to decode it with multiple passes ... We were once expected to approach texts for information. We would forage until we found something of interest, become still and read."

Do that now. Become still and read.

This terrain is to be explored, sampled, revisited. The context for these words will be revealed over time, shifting in shadow and revealed in the bright light of inspiration. Its discovery is a personal delight.

This book is a considered collection of ideas, proposals, observations. "Information" is too narrow a term for its content. Selected from the output of a quarter-century, the essays are works of theory but also of imagination: outrageous proposals and intimate reflections, vehement commentary and close readings. You might even discover fragments of your own memories here.

In reading this book, you will come to know something of Tom Sherman. He will tell you stories, some of them reflections of his personal experience. But, of course, such "information" is seldom transparent. Some facts will be buried, or camouflaged, by other items of interest or argument overlaid or tucked alongside. You will meet more than one Tom Sherman in these pages, some of them the artist's constructions as performer, conduit, researcher, commentator. One facet may become another, or its opposite.

Tom Sherman is an artist who has drawn deeply on his study of communications, the information economy, and natural science. His writing works within and between these disciplines, postulating new ideas and congruencies, revealing possible truths for our future. In some cases, his speculation has become fact.

These writings are narrative literature as well as performative document. They offer parallel lives as versions of reality or "truth." There is more than one "I" acting here, whether as assumed identity, as "cover," or as alter ego. The Tom Sherman speaking in "Shades of Catatonia" differs from the author of "Wordwatch Systems," where neither is a simple facade. Nor are these characters like the writer proposing "More Dead Artists" as a solution to Canada's lack of history and proliferation of unassimilated young talent. As he points out in "Once Living: In a Healthy State of Paranoia," "Don't I owe it to myself to keep my eyes open for an ambush?"

Read carefully. Be open but not gullible. Be challenged to take each word at face value, but also as metaphor, analogy, even as contradiction. These ideas and proposals are history as well as future: possible options or states to be avoided. Or all of the above. This book is a work of substance, the result of research, assessing and valuing the information economy and international intellectual community. It is also mischievous, a site where what Sherman has described ironically as "raw personal material" (in "Millennial Spurn") is less a waste product or virus than a subtle initiative, a recommitment to actual "real" life with all its impurities.

The book embraces the natural world, yet it values the assistance and company of machines, while cautioning against a blind acceptance of business bromides and directives. *Before and After the I-Bomb* is a complex work, full in fact of raw personal material but also brimming with intellectual rigour and original insight.

TORONTO, CANADA, OCTOBER 2001

Prologue

I write every day, usually in the morning. I fill notebooks and hard drives wherever I go. It's a habit I picked up in the early 1970s, when I found it necessary to articulate the ideas central to my work in installation, performance, video, and audio art. First, I wrote to develop ideas I wanted to pursue in other media. It wasn't long before I convinced myself that language was a medium worthy of full attention. After all, conceptual artists such as Lawrence Weiner, Sol LeWitt, and Douglas Huebler had already established in the 1960s the essential role of writing in art. But I had to find my way of working with the written word. I discovered it's the direct application of ideas that is attractive. Language is where the mind hits the pavement.

Writing for me is an act of discovery, and where things originate is often much more interesting than where they end up. Forming texts, objects of thought, has always been my preferred manner of making things. There is nothing wrong with forming objects from physical materials, so long as the materials don't obscure the thinking behind the work. I've always found media like video and performance so beautifully transparent, because television and the "life is art" mantra have made these media common and direct in the best sense.

I need to write. Not only do I establish and maintain my connection with the world through acts of description and the examination of accumulated observations, but I also construct continuity and purpose in my life by determining the next target for investigation using aesthetics, rules intrinsic to the creative process, as a guide. My obsessions are generated by the process of structuring accumulated observation. My profession is that of the visual artist who writes, often as an end in itself. I'm not a novelist or a writer of short stories. I've been told my prose has literary dimensions, but I'm coming from a different tradition and place. I discover what interests me and set up the methodology for following the thread from one idea or image to the next — laying out the work for the next day and the day after. It's the process of working and the way it absorbs me that I find satisfying.

This book is an accumulation of this process of discovery over time. It is also a personal history of the late twentieth century. As the personal history of someone concerned with how the story of technology and civilization will unfold, the book is intended to be an instrument for negotiating the dense cloud of scrambled, technobabble that obscures this period of transition. For a long time, we held clear, absurdly utopian images of the future. The future was something to look forward to. Technology was at the heart of the new, improved future up ahead. This book reveals another perspective on the future. The future seems to have reversed directions and is collapsing upon us. With developments in neuroscience, genetic engineering, and networked intelligence, we find it difficult to cope with the present, let alone the future. Confusion reigns. We're going backward, side to side, and forward at the same time. There is no collective idea of where we are headed. The future is multidirectional. With no collective vision, the individual is at the centre of the universe again.

My perspective, though global in scope, is North American in reality. I was born and raised in the United States, living until the age of twenty-four in Michigan. Disillusioned with the American culture of Vietnam and Watergate, I emigrated to Toronto in 1971, where I fell in love with the ideals of Pierre Elliott Trudeau's Canada. I spent the next twenty years in Toronto, Ottawa, and rural Nova Scotia, returning infrequently to the States, usually to visit my parents. In 1991, I was not-so-gently prompted by a severe recession in Canada to take an academic position at Syracuse University, where I became half-American again, residing and working in upstate New York during the academic year. For the past decade, I have migrated back and forth from Syracuse to my home on The South Shore of Nova Scotia, from where I write these words. The texts throughout this book oscillate in point of view depending on where I was (the actual ground) when writing.

This selection of texts tells the story of an artist perceiving the world through media. There is nothing exotic about my experience; a media-saturated world is now available for all in the developed world to share. For whatever reason, I chose to be "wired" before telecommunications technology became the necessary "option" it has become today. I've always found it preferable to negotiate reality with instruments. Partly this preference was due to immersing myself at an impressionable age in McLuhan's Toronto, but more fundamentally it probably resulted from my reading massive quantities of science fiction as a boy and playing with short-wave radios and tape recorders.

During the summers of my childhood, I also spent a good part of every day observing and collecting insects, and marvelling at their strength, diversity, fecundity, and seemingly supernatural abilities to sense the world and communicate. I was a naturalist then, and I remain one now: when observing and describing the world, tell it like it is. Describe the good and the bad, and leave nothing out because it is disappointing or frightening or unlikely. In these texts, I've tried to honour that ideal.

I have entitled this book *Before and After the I-Bomb* partly because I've had the reoccurring thought that we have been overrun by our technological inventions. The growth of technology was rapid throughout the twentieth century, but around 1994 or 1995 there was a flash, as if the world "blanked" from an overdose of telecommunications. I remember looking around and seeing that few people seemed to notice. The new had become the same old, same old. There had been a period of slow burn since the mid-twentieth century, during which time telecom-reality gradually eclipsed plain-old "primate" life [in North America]. Up until the mid-nineties flash-point, our use of television, video, computers, and telephony had mushroomed, but we were still managing to evolve our behaviour to cope with, and even enjoy, the imposition of mediating devices.

After the I-bomb exploded in the mid-nineties, there was a sudden challenge presented by digital, networked, global telecommunications. Against the wicked speed and inhuman precision of digital-telecom, we look more and more like primates: simple creatures content with routines better suited to vanishing domains. Too few of us noticed the flash point, the moment when we were cast as biological anachronisms. Most of us simply snapped, unable to make the appropriate changes to accommodate the explosion of new technology. It was a cultural, evolutionary breakdown.

Before and After the I-Bomb is a detailed account of the slow burn of telecommunications through the late twentieth century, and into and following the flash point I described above. Now that the future has reversed directions and imploded, it's time to take stock and move on, perhaps down roads already traveled. I offer these notes as proof of what happened, and as a travel-guide for tomorrow.

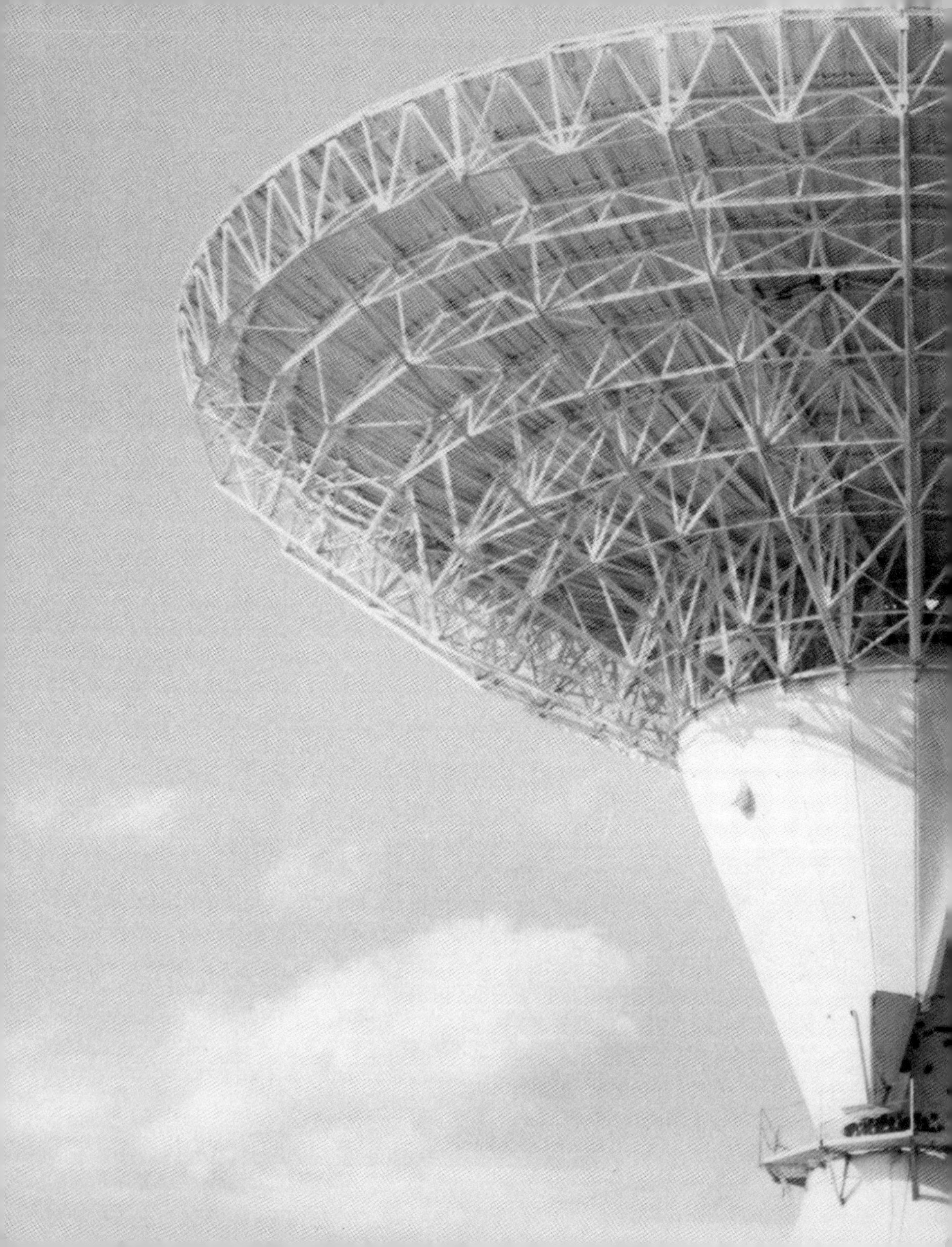

CHALLENGES > 1

TOM SHERMAN

After the I-Bomb 1999

There is a lot of anxiety today about information overload, corporate control of the news, and commercial propaganda flooding the expanding global technological infrastructure. People seem to feel they are coping, but they fear being overwhelmed in the near future as networks increase their saturation of mediated images, sounds, stories — a plethora of calculated manipulations.

The information bomb exploded in the early 1990s, when computer networks attained a critical speed and scale, flipping the gates wide open to unleash a torrent of blinding, deafening code — a thunderous explosion of advertising, entertainment, voice, and data. Now, we must negotiate continuous, relentless input. We have become the organic components of an integrated global data and information system.

This media pressure keeps us afloat, even as it blows through us at the speed of light. Our existence, who we think we are and how other people see us, is defined by the shadow we cast. We are exposed, as if by x-rays, by the increasingly violent light of the I-bomb. Like the moth fascinated by the flame, we can't resist the light. This fatal attraction to media keeps us coming back for more. We are gluttons for massive doses of symbolic code. When we close our eyes or turn off our info-appliances, we mull over the afterimages, the disembodied voices, the imprinted rhythms. We are eternally plugged in. Our equilibrium is skewed, buffeted by the message storm.

Faced with this assault, it would probably be wise to wear welding goggles and earplugs. On the contrary, huge numbers of individuals are wiring their homes with high-speed network connections, buying industrial-strength information processors, setting up home theatres with high-definition screens and surround sound, pointing satellite dishes toward the heavens, taking cell phones to bed at night. The symptoms of information addiction are not subtle.

Most thinking people believe that emotional and intellectual survival depends on a managed exposure to information overload. Surviving an exposure to excessive levels of media information can result in a strong sense of media literacy. But this managed-exposure approach is developing an immunity the hard way.

Interactivity is required to become media literate, and it is highly recommended for counteracting the numbing effects of the I-bomb. Literacy is not just sorting out and digesting media information, it is also learning how and when to issue messages, casting them into the information torrent and altering the immediate media environment. The danger with interactivity, however, is that individual message sources are often targeted for response by highly focused, industrial-strength, corporate or government media campaigns. More advertising and propaganda are fired directly at individuals who care to interact. Filters and shields (such as advertising zappers or intruder-specific firewalls) are key to surviving this return fire. Such protective devices are necessary for those choosing to refine their media literacy through interactivity.

TOM SHERMAN

Blanking 1996

A new psychological phenomenon is emerging in this era of hypertelemedia, information abundance, and information overload. People of all ages and walks of life are blanking; that is, they are shutting down or experiencing momentary ruptures of consciousness or, in very severe cases, blanking sometimes lasting for days. Blanking is not attention deficit disorder (ADD) or daydreaming (dd), but a breakdown of consciousness brought about by sensory and cognitive overextension induced by hyperconnectivity.

These days, we rarely choose to focus on one coherent stream of information. Rather we simultaneously gather data from multiple sources. Instead of simply listening to the radio or watching TV, we listen to the radio *and* watch TV and read a newspaper, magazine, or book, while we have something to eat, while we have a conversation on the phone, while we stroke our dog's tummy with our bare foot. This is how we function in leisure time: we choose to compose our immediate information environment from multiple sources, mixing our multilayered reality on the spot.

We've become good at packing more and more sources of data and information into shorter and shorter days. In school or the workplace, we have learned to concentrate on specific required tasks in chaotic environments characterized by high levels of noise. We are always in training, learning to swim in more and more confusing, turbulent currents of data. Successive generations have become better at coping with louder, busier, more crowded, more discordant surroundings. When we come home to a dark, quiet apartment, we need to flip switch after switch, turning on info-appliances until we achieve our preferred quota of information. We panic without this pressure, a media presence that is akin to atmospheric pressure. We sleep as an escape from exhaustion, but with our televisions and radios and computers left on.

Many of us have multimedia workstations to turn up the volume day and night. Many work at home, with our workstations transformed into playstations and vice versa by simply pointing the machines in different directions. Perhaps more accurately, the difference between work and play is determined internally, psychologically, as we adopt these different attitudes. We mix our multichannel, multidimensional realities, in a private place, wet and cool and blue with bioelectrical equilibrium.

Hyperconnectivity is the buzz. When we're flying high, we tie into anything and everything, and we enjoy building bridges between audio and video and data, and touching and looking and tasting and being here and there, while wanting to be somewhere else, alone together in sweet intensity. We're strong and immense and spread out in tandem with the forces of the universe.

But then we blank, and it's frightening to go down, to crash without sound or punctuation. To terminate into emptiness. First, it's scary, and then it's disappointing, an experience racked with futility. (A loved one strokes our hair in the distance.) Our fingertips are frozen in the curl of our empty hands. Emerging from blanking you realize that overload has been achieved. You've gone farther than ever before. You have pushed yourself into the organic reset mode. You'll have to stay quiet or disconnected for a while, but step by step, you'll begin to reload again. Then you'll progress as if dancing through a series of inevitable contacts; tying yourself into this and that until you feel whole again; up and running, functioning, expanding, turning up the volume, finding the rhythms, opening your eyes, wider, stepping out farther, reaching across, bridging impossible distances and differences, until again you blank.

TOM SHERMAN

Waiting for the I-Bomb 1999

We seem to be waiting for the information bomb, expecting it will be dropped sometime in the twenty-first century. There will come a time when the world will be perilously unstable, a time when the I-bomb will be needed to wipe, stabilize, and reset the integrated global data and information system.

The I-bomb's elaborate infrastructure has been in development since the mid-twentieth century. Piece after piece has been assembled and implemented, rationalized by sweeping statements of public benefit. Television, satellites, cable, stereophonic sound, video, VCRs, personal computers, CDs, fax, the Internet, fibre optic cable, wireless phones, the Web, and digital everything — all these things, we are told, are all good for us. Real technological advances. This complex and all-pervasive media infrastructure has grown like a garden during a warm, moist summer. We now have more tomatoes than we can give away.

Governments struggle to regulate this garden of media technologies and to maintain the right and expertise to control this infrastructure in a global emergency. Commercial enterprises need little encouragement to spread out and overwhelm the system and its inhabitants with calculated manipulations of endless permutation but identical intent: to trick people and take their money. Common folks are invited to participate from the start, fleshing out the new territory from the ground up, while subjecting themselves to an atmosphere of increasing vulnerability and manipulation.

Each new technology creates a new territory, offering opportunities for social exchange and economic development. For instance, in the mid-1990s, the early days of the World Wide Web, there was unlimited space for everyone. The early adopters were able to settle in and trash the new environment. With no expectations or restrictions people set up sites that tended to sprawl and led visitors into dead-ends and frozen screens. Then, business interests developed and rents began to rise. There are always great spaces with low rents at the onset of gentrification, but once a territory is coveted by commercial interest, the exploitation may commence.

Nothing is more valuable than a good address in a territory created by a new technology. This is true when things are going well and it remains true as conditions change for the worse, as they have on the Web. When things get tough, there is nothing more important than a good address.

TOM SHERMAN

Shades of Catatonia 1999

Sometimes when I'm watching television, I think about driving. Watching television and driving feel the same to me. Sometimes when I watch television, I think about animals. A couple of weeks ago, I was driving on a highway, a divided highway in upstate New York, when I came around the bend. There was something yellow off to the side of the road up ahead. It was golden. As I came upon it, I realized it was a golden lab, a large domestic dog. It was tearing into the insides of a dead deer. It was ripping into roadkill on the side of the road. I don't recommend that people watch television when they're driving; it's hard to stay focused on the road. But I like to think about driving and animals when I watch TV.

When I watch late at night, I have trouble unplugging. I have trouble withdrawing. I have trouble killing my television. I get so heavy in the chair. I feel like I'm sinking at the end of the day. I feel depressed. I sink so deep. I slide under. I slide way under and then crawl into bed. I get dead tired, but I don't sleep very well. When I get like this, in the morning I try to take another tack. I plug myself back in. I go electronic. I get into my technology and feel the surge. I go straight up like a rocket. I get high. I get manic. I get nervous and edgy. I get wired. I go straight up with my gear. Give me more information. Give me a massive fix. I love the surge. I get manic and then I crash, and I slip and slide under. It doesn't matter that I go up and down like a yo-yo. I love the ride. I'm comfortable being out of kilter. I don't mind losing my balance. I accept my limitations. It's hard to get a handle. I never have enough time to think it out. I just turn it on and ride. Getting lost to find

myself, out of kilter.

That's why I pay attention to the experts. I listen because they try to tell me why. Why I'm being jerked around. Why I'm living on a roller coaster. Why I've got a bipolar disorder. The experts. They're the spokespersons for the same condition on a global scale. They're fleshing out a world in conflict for everyone to see. They're diametrically opposed. On the one side, you have the technofreaks. They're the utopians. They're cool and crazy and oh so positive. They're on a rocket straight up. They say go with the flow. Get some new gear. Get your feed. Extend your reach. Become superhuman. They're the globalists. They're the libertarians. They love the machines.

On the other side, there's the resistance. Those who choose to put both feet on the brakes. These people are the social realists, the reformists. The champions of those left behind or soon to be abandoned. They describe themselves as the critical thinkers. They're the people who mock the nerds and hassle the carefree. They spout doom and gloom. They are hard to ignore. These are the experts, rendered in black and white. They speak for both sides and have fragmented, conflicting visions of where we're heading. The experts form a societal opposition, a global division. They highlight the conflicts, but they offer no solutions.

Most reasonable people can identify with both sides. It's a balancing act. Most people have learned how to deal with conflict, with disorder, with paradox, with a conflict of orders.

Most people have learned how to cope. They've suffered losses, they've flip-flopped on important issues, but they've survived. They've taken sides, and they've been burned. They've lost face. They've had to leave their bodies behind. Their nervous systems are not so central anymore. They're overextended, they're out there taking hits, but they hold enough back so they can still feel inside. They still manage to maintain control. That's what it's all about. Control. Controlling your emotions. Make up your mind to take control. Make up your own mind.

I worry about losing my sense of self. Sometimes I don't feel like myself anymore. Maybe all my thoughts have been implanted? I wonder if any of my thoughts are my own. So many people are trying to persuade me and influence me and make me buy this or that. Most of the time, I feel there's nothing left that's mine inside. I can secure myself with things in the material world, but I get that sinking feeling. On the rare occasion I have an original idea. But then, before I can act, I see my idea on some screen for everyone to see. Someone is grabbing my thoughts. My ideas are being sucked right out of me. It's like thought-extraction. It makes me paranoid. But I don't panic. I just take a deep breath and relax and try to re-centre. I wait for friendly instructions, commands, or prompts. I've learned to look outside to find myself. I feel a kind of warmth right now watching the incoming information on my screen. I feel the impact of distant collisions deep inside, and it's a feeling of information passing through me, and there's a subtle impedance at the site of this interference. That's me inside. I can feel myself. I find myself, in the grip of a manageable level of thought disorder.

It's hard to recognize myself when it's faint or transparent, when it's perforated with millions of tiny holes. All I know is there's a conflict of orders in my mind. I want to make up my own mind. I want to feel the information passing through me. I want to be able to resist. I feel I could resist if I could just get back far enough, with enough distance and perspective to figure out what's going on. But the ebb and flow, the implantation and extraction, are devastating, and the instructions, commands, and prompts are inescapable and sadly necessary. They're necessary for survival. The incoming signals define us. I'm starting to think the way the announcers think. They know. I don't know. They know. I don't know. There's got to be a way of avoiding the dominance of this programming. There's got to be a way of ducking under the incoming waves of commands and instructions — the code. Wave after massive wave of symbolic code. There must be a way of ducking under it all.

> > >

We live in a universe that's psychologically subversive. Everyone is trying to plug something. The screens are the marketplace and the battleground. People are drawn to the little screen for entertainment, information, and social integration. People go to school on the screen. This is where we learn things; it's the site of continuous learning. The people behind the screen are always shifting its emphasis. It used to be that education was political. We would learn the things we needed to know so we could participate in our societies politically. We learned so we wouldn't be overwhelmed. Life is painful and dangerous sometimes. Problems get in our way. The way you feel and what you can actually do about it is the essence of politics. You can learn how to make up your own mind. If you feel you have the ability to link personal problems with societal situations and broader issues, and to translate these relationships into political actions, you can manage your world. That's the way schools used to operate. Education was political. But we've abandoned the political aspect of education. People have decided that education is no longer political.

Education is now about training. Not just pure manipulation and conditioning, but training. Education now prepares people for better jobs. It's manipulation and conditioning with a projected financially beneficial outcome. Although the financial goals of education are clear, the political benefits of education have been skewed. The people behind the schools and screens have decided to actively manage your emotional landscape. The main political goal is psychological and social cohesiveness. Engineering common or shared feelings. I used to feel special. That's the way we all used to feel. I used to feel I could sit in a room like this one, and only I would be here in my unique way. No one else could see the way I see. Things felt different to me. I used to think that individuals saw things differently. But now everyone sees the same. We share a sameness. We submit to this sameness willingly. It's a seduction. It's a persuasion. Sit back, relax, the pattern has motion, a nice pace, and you are in control of what you watch.

If you believe that, you'll believe anything.

There's a lot of training going on with all this new technology. If you're wired on-line, you probably feel you are master of the technological universe. If you're wired, you're probably interactive. You're cool. You've got your hands on the steering wheel; your foot on the accelerator of a dream machine. You can go where no one has gone before. You're loaded and immersed, and you're fearless. You are participating in the creation of your world. You are in training. Make up your own mind. Feel the power. This feeling is where it all starts, with you in control.

It would be nice if someone would teach us how to tell the difference between propaganda and education. Education is at least the reinforcement of traditional values. We can rely on what our parents and teachers have told us. It feels right. It feels good. But it would be very useful if we could tell the difference between propaganda and education when we're out on our own. I don't mean that crap about dialogue — the "back and forth is always better" angle. I mean knowing the difference between propaganda and education, so we could feel excited about learning again, so we wouldn't have to be afraid. Driving defensively or, worse, driving with the brakes on — it's hard to feel good about learning. Life is tedious when we're always playing it safe.

There's nothing new on the screens. It's just a pattern that changes in density and frequency. It's just a story, a series of overlapping stories. They overlap and recycle, and become redundant. If you're trying to tell a story of any depth or complexity, all you have to do is repeat yourself. If you want your story to hold together, just use redundancy. Vary the details and pace, but repeat yourself. People of all ages like to hear the same story over and over again. It makes us feel secure.

During the Cold War, people used to talk about propaganda. Evil forces were trying to brainwash us. That was long ago, back in the 1950s. People were beginning to go to school on the screen. They used to tell the school kids a story. If they drop an A-bomb in your neighbourhood, you can take shelter under your desk. You'll be safe there. This story was an exercise in emotional management. Panic management. I don't think anyone believed the children would be safe. But it felt good to have a plan. That's one of the benefits of education. It is comforting to share the same story with everybody else, whether it makes any sense or not. The story these days is if you go to school and study hard you will prepare yourself for a better job. That job will likely be connected with new technologies. If you learn how to interact with computers, to become technologically enabled, you will be a survivor. You'll survive the digital tidal wave. It's a new twist on emotional management. Panic management. If education was once political, and now it's all about training, I wonder what they'll do with education next.

I feel a need to go under, to duck right underneath. When I'm staring into the pattern of a screen, I feel like a deer in the headlights. I'm frozen in the light. I may be temporarily incapacitated, but I'm more resourceful than you think. I'm transfixed, but I have a plan for survival. When the density and frequency of the stream of symbolic code is too much, we are going to have to duck under. We are going to have to go subsymbolic. If we are going to survive our overexposure to the crushing density of symbolic code, we are going to have to slip under the mounting pressure. We will have to find a way to withdraw, to get back, to create some distance and some room for contemplation. Don't panic. Overload equals pattern recognition. Just hold your ground. Check out the space between the images. Check out the space to the side, at the periphery. All you have to do is relax, absorb the flow, and slip under. Just relax; let the pattern emerge. It's got a nice pace. It's designed for you. You're special. You are in control of what you watch.

I feel a need for a change. I need to get out of the line of fire. I feel like I'm spreading myself thinner and thinner. I'm still absorbing energy. I'm taking hits somewhere in outer space. I'm absorbing this remote contact, this distanced sense of interference, right into my guts. It feels good. So there's a way to make it happen inside. They talk about the dangers of roboticism. Command automatism. We'll all be robots in the future — totally programmed organic robots. We'll all be thinking on the right sides of our brains. We'll be conscious without being self-conscious. It's not so bad to make the big slide over to the right side. It will make us feel secure. Home is where you hang your head. If you can find a place that feels good, you're more than halfway there. You're going to have to slide over to the right side, to duck under, to go subsymbolic. You're going to have to. I feel like a deer in the headlights.

> > >

There's a way to go under the shower of empty images and sounds. It's an escape from confusion and pain. There's a conflict of orders in my mind, but there's a way out. I start to feel bad when I'm on-line all the time. I feel like

I'm living in a world where the good things have been left behind. It's painful to sit in front of a computer on a day when the sun burns brightly and the earth smells sweet; on a day when you could naturally slide under and go subsymbolic. When you're going under, there's a realization, a peace of mind. It comes from a strange place, deep inside. You feel like an animal. We have this human conceit. We make the whole world human, in the worst way. We pile on the symbols, increasing the density and mass of the human conceit. This human conceit, I call the human imposition. The human imposition fucks with our natural disposition.

There are psychologists who study animals. There are doctors of all kinds of minds. Doctors of creatures of all kinds. There are bird doctors. Let's jump from the mammals all the way back through the fish and the reptiles to the birds. Let's go back to a deceptively simpler time. The birds are smart. They've got brains that are bigger for their body weight than those of mammals. And they already think on the right side. When animal psychologists try to prove that animals are conscious, that animals have conscious thoughts, they wire them up and read their brainwaves. The bird doctors believe that birds think subsymbolically. Birds are conscious and aware of how they feel, but they don't show a range in the symbols they exchange. They think with the right side naturally. It's the way they're wired. We're different. We're hardwired to spend time on the left side with our languages, our crazy symbolic playground, and our human conceit. But we, too, are sliding over to the right side. We're being forced to. Overload equals pattern recognition. We're going subsymbolic; we're going under, by necessity.

So there's hope for us and, by implication, for the other mammals and the fish, reptiles, and birds. I had the strangest thought the other day. I realized I'm starting to think like an animal, like another species. I'm starting to think like a bird. Scientists think that birds only talk about two things. Birds don't have thousands of stories to tell, nor do they spend time making small talk. They don't need an elaborate vocabulary because their messages are simple. They cut to the quick. If you wake up at daybreak, say five in the morning, and birds are out there talking, it is the same kind of talking they do later at dusk, in the twilight hour, just before the sun goes down. The bird doctors think that birds say only two things, morning and night. They say: I've got a great address. I really like where I live. It's a really great tree. It's home. Stay away. It's my tree. I really like my home. It's a good address. I really like my place. And the other thing they say is: Come on over. Let's have some sex.

That's what they think the birds talk about. They say: I've got a great address. I love my place. Want to come over and have some sex? Make up your own mind. It's simple. I've got a great address. I love my home. Home is where I hang my head. Hey, come on over. Let's have some sex. Come on, make up your own mind. It's simple. I've got a really good address. Come on over, have some sex. We can learn from the birds. You can learn to let go, to slide right under. There's hope. You don't have to worry about becoming a robot and losing consciousness. We're all going to be connected again, big time. You've just got to learn to think with the right side. It's easy. Hey, come on over to my place. It's a great address. Home is where I hang my mind. Come on over. You can make up your own mind. Let's have some sex, come on. Just lean to the right.

I've got a great address. People visit me. They love my place. They come on over and have some sex. Make up your own mind. The pattern has motion, a nice pace. Just sit back, relax. You are in control. Make up your own mind. Because home is where you hang your head. It's where you live. It's where you reside. It's how you feel inside. I've got a great address. Let's have some sex. My place is your place. My nervous system is not so central anymore. But I can still feel deep inside. It's simple. I can make up my own mind.

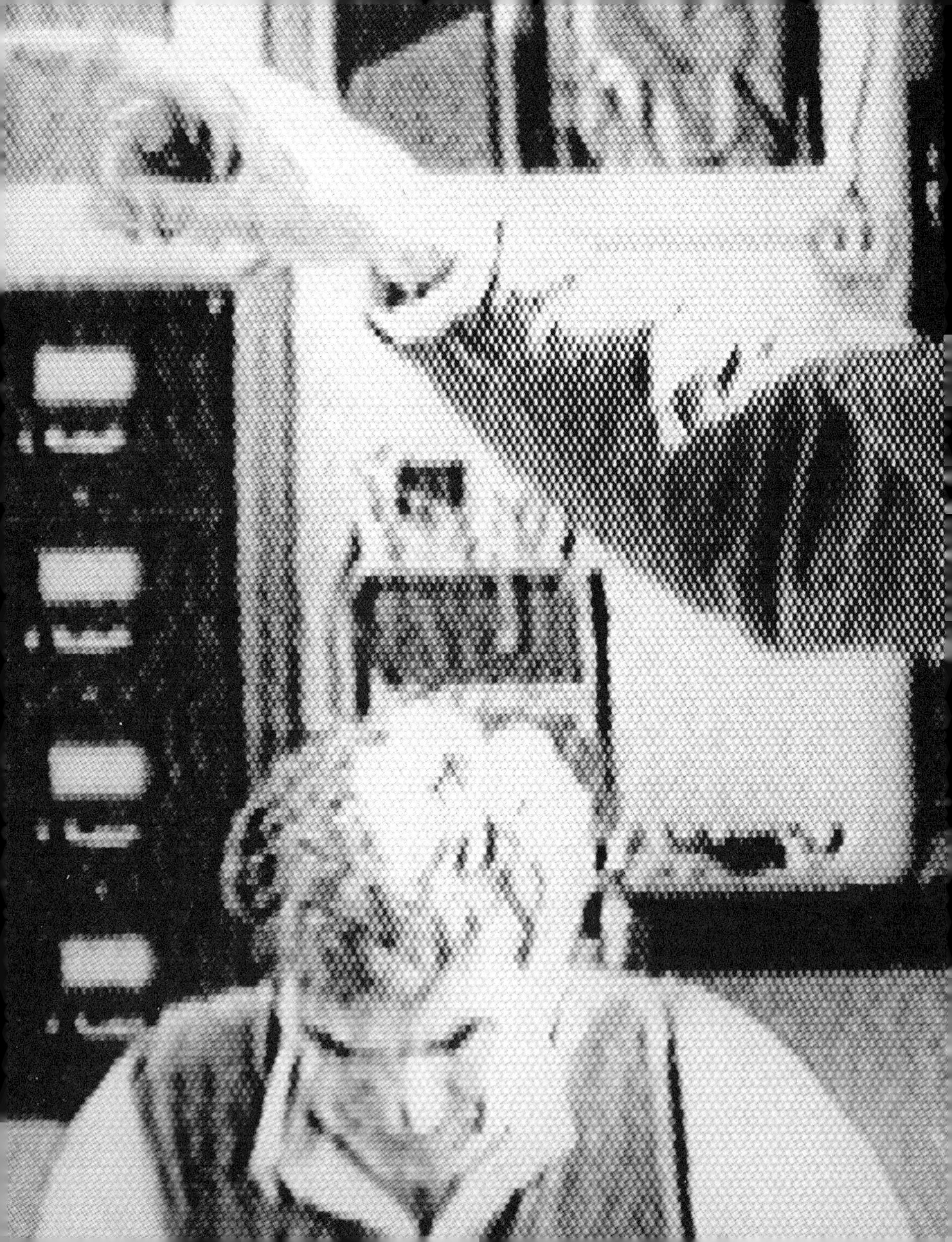

THE CHANGING WORKPLACE

TOM SHERMAN

Wordwatch Systems Inc. Watches What You Say 1995

Syracuse, New York: February 5, 1995

WordWatch Systems Inc., an information-age consulting firm, is using computer voice-recognition technology to study verbal communication in the corporate sector, analyzing spoken-language artifacts to predict trends in behaviour and ultimately to anticipate changes in the goals and performance of a wide range of corporate executives and information specialists.

WordWatch, a consulting firm in its fourth year of operation in Syracuse, New York, has been contracted to install voice-recognition and transcription systems in the boardrooms of several seemingly unrelated businesses in Central New York. WordWatch guarantees the confidentiality of all business conversations. Its automated-listening systems monitor and record sensitive business conversations in a quest for linguistic statistical information.

Tom Sherman, president and chief executive officer of WordWatch Systems, explains what his firm is looking for when it analyzes recordings of corporate conversations: "Our systems record and transcribe everything that people say; we record literally millions of words in a specific boardroom over a four- or five-month period. We can then perform comprehensive, painstaking analysis of linguistic patterns as they develop. We use expert systems to process the tremendous volume of language we collect. In our most recent contract research, we are looking for patterns in how abstract nouns are formed and used. In particular, we are interested in words that end in *-ity*.

"Words like *interactivity, compatibility,* or *creativity* are very hot in the corporate world today. These 'ity-words' are abstract nouns, where verbs or adjectives are transformed into nouns through the use of the suffix *-ity*. These resulting abstract nouns describe states or conditions affecting work, problem solving, marketplaces, valuable processes — you name it."

Sherman is particularly interested in how these ity-words express values. "We are conducting our most comprehensive research project to date, monitoring the boardrooms of a number of diverse organizations. What people say (and how they say it) in a small design firm, or a major insurance company, or an electronic equipment manufacturer differs linguistically. It is possible to see trends in language emerging, growing, and moving from one context to another, from one business sector to the next. For instance, the word *creativity* has been growing in popularity across all economic sectors as the volatility of time-sensitive information takes its toll on the stability of various markets."

Although Tom Sherman is unwilling to violate the confidentiality and proprietary nature of WordWatch's contract studies, he is willing to give a few hints about what he sees in the statistical linguistic indicators developed by WordWatch. "Ity-word use is increasing dramatically, a trend that expresses a general shift in economies globally. The harsh realities of global competition, diminishing resources, shrinking profit margins, and environmental concerns have forced many businesses to market processes as products, as they abandon the production of material goods. Today, more and more businesses are selling ideas, not things. Ideas are generated through conversations. Boardrooms are typically the sites for a lot of wordplay. People get bored in 'bored rooms,' and predictably they tend to improvise through speech to give their new ideas visibility and profile.

"In the last quarter of 1994 and the first few months of 1995, the number and frequency of the use of ity-words is up 63 per cent from comparable studies of a year ago. At seven monitored sites in Central New York, we have identified 371 ity-words in use, some of which are being used thousands of times a month. These ity-words express the values of corporate executives and information specialists of all descriptions. Ultimately, ity-words help us identify what our clients think is important. In the oral proceedings of our corporate business culture, the values we express are always connected to ways to make money. At WordWatch we are in the business of helping people find new ways to make money. We do this by listening very carefully to what they say."

TBDF (Transborder Data Flow) 1982

THIS ESSAY IS WRITTEN IN THE VOICE OF A FICTIONAL CHARACTER, ANDREW CZESZAK.

When you find your dollar isn't worth the paper it's printed on, it's time to start determining value under a new set of rules. My dollar says the times are changing. With everything changing so rapidly, I'd say the criteria for establishing value are shifting; shifting like a load that's not tied down. Money alone won't make you secure. Don't misread me: money doesn't hurt those who have it. But money by itself isn't enough. You've got to have someone. You've got to belong.

I'm living here in Canada. It's where I belong with my wife, my kids, and my job of six years and a couple of months. It's cold here in Ottawa, but that will change. I make pretty good money. I'm a public servant working in a high-rise nine to five. I measure TBDF — transborder data flow — for the DOC — Department of Communications. I read machine language all day long. We've got a bank of mainframe computers sitting in the basement. These ancient Honeywells are automatically fed by manned minicomputers (IBMs) in 211 remote DOC checkpoints interspersed across the borders of the provinces and territories. Just like the money banks, we have to keep the "books" on our balance of payments.

I've got a decent view from my office. Actually, I manage a superport. Virtually everything that moves is compiled on the nineteenth floor and cabled up to me. My desk is on the twenty-sixth floor of 44 Kent. I have hard copy just as soon as the fresh numbers arrive via VDT — video display terminal. Christ, what am I doing wasting your valuable time talking shop? The fact of the matter is TBDF doesn't mean a hell of a lot outside our department. The most important thing for me to communicate is the reason we are recording and analyzing TBDF. At the date of this writing, there is a general fear in the federal government that leads us to.implement our current program of data flow surveillance. We are deeply concerned with setting up security for the eventual preservation of your electronic cultural heritage.

Chances are you didn't even know your electronic cultural heritage was being threatened. I can sympathize with you. If I wasn't occupying this desk, I would be hard pressed to pick up on the other end of this conversation. Let me elaborate. TBDF is not a new concept. Every time a book or a newspaper or a telegram is sent from Canada to the U.S., or vice versa, that's Canadian-American TBDF. Phone calls, radio and television programming, exchange of video and audio cassettes, high-speed digital communications — all terrestrial or satellite communications between nations make up the data exchange described as TBDF. As you might imagine, because Canada has a relatively small population with a very sophisticated communications system, we import far more data than we export. Speaking generally of the Canadian-American information exchange, Canadians are the carriers of American content. The Canadian broadcasting and cable industries resemble insect vectors as they inject American electronic culture into the veins of the Canadian people, feeding a life-blood addiction as most Canadians feverishly subject themselves to massive doses of American media.

A few people hate to see so much American television flooding the country because they realize it cripples the Canadian television industry. Indigenous television production is curtailed before it gets off the ground by lack of demand in the domestic marketplace. Without a solid domestic audience and an advertising or subscriber base, how can the undernourished Canadian cultural product compete? In this day and age, when almost every country is sucking hard on a one-way flow of information from the U.S., how can we expect to participate in an international dialogue if we cannot increase the value of our own information? Television is the most visible indicator of our unhealthy situation. We're watching theirs, but they're not watching ours. We are not even watching our own.

My wife, Eleanor, and kids, Michael and Angela, survive nicely without television in our home. Eleanor and I decided to get rid of our TV during Michael's infancy. We noticed his classic response to the light. He's fifteen now, so we haven't had a television in our home since 1967. This year he has begun to complain bitterly about the situation. He refers constantly to our "missing TV." Angela doesn't seem to care if we have a TV or not. She is eleven and loves to read. Both our children have excellent reading skills, and they do very well in school. Michael's grades have been especially good this year. He manages to log about four hours a day in front of a VDT. Half his classes are computer-aided, there are a couple of dozen Apple II's linked up at Alta Vista Collegiate. I've heard the school is upgrading to Nabu's new micro in June.

It is 1982. Computers are everywhere. I want my kids to know how to use computing technology. Television is already half of our environment. I want it to continue to be the external environment. I refuse to allow it to explode the confines of our living quarters. I've seen horrifying statistics compiled by our Arts and Culture branch. These studies indicate that television viewing will eventually lead to a total fragmentation of the family unit. Just as the fragmentation of the audience is occurring on a national and international level, we may also observe the fragmentation of the family under the stressful influences of television. How many times have you see an argument develop between family members over program selection? Authority in today's average home is literally defined by who controls the channel selector. Of course, the solutions to such problems are multiple receivers, all providing access to the same field of information. In our family, we would have to own four sets, all colour, all hooked up to cable, to keep everyone happy. We can't afford four colour TVs, nor do I want to see our family living in four different corners of our home.

Even though I may seem unreasonably negative when it comes to television in our home, I am not generally antitelevision. Video is an amazing medium, and it will soon be inseparable from computing technology, just as film is presently indistinguishable from television and video. All of the advanced information technologies will combine by the turn of the century, and eventually our entire recorded history will be available in an electronic form. I don't think our daughter, Angela, will ever be all that enthusiastic about the prospect of an electronic cultural heritage. Eleanor and I would like to see her mature in an entirely different direction. Contrary to the masses, she will grow to possess rare qualities of memory and imagination. Our son, Michael, already leans the other way. It is natural for him to strive to record the present. With a keen sense of play, he is constantly assembling all aspects of his sociological, economical, and political reality. When the time comes for him to reject our support, he will gravitate professionally toward the practice of high-level information processing. He will be drawn like a moth to the power of information. I know Michael very well. My son will never be able to resist the promise of intensive illumination.

When I walk to work in the morning, I pass by the National Defence Buildings and the old National Gallery. It takes me twenty-one minutes to walk to work at what I consider to be a brisk clip. During my walk, I concentrate for approximately eight to ten minutes on developing a structure for thinking that will permit me to formulate the characteristics of an electronic cultural heritage for Canada. I carry this mental activity into every working day. I confine myself to the identification and memorization of imagined parameters. My intention is to invent a formal structure that may allow for the visualization of elaborate cultural constructs. I find great reward in being the first to recognize a complex idea in the full beauty of its holistic form. The initial development and immediate maintenance of an electronic cultural heritage is well worth any conceivable outlay of public funds.

First, big money will have to be sunk into the development and manufacture of quality hardware. Then, once this hardware is in place and we reach the state where our massive memory banks are known to be empty and prime for investment, we must set up basic editorial controls. We must have solid regulatory protection in place from day one to ensure that our banks are not flooded with an overflow of information from the United States. Along with well-designed regulatory protection, we must offer attractive incentives to encourage Canadian information producers to deposit their information at home. Initially, we must stimulate Canadian information investment through full-spectrum seeding until a clear industrial strategy can be extracted from the morphological characteristics of the investment base. The preliminary stratification of the information base should clearly indicate peak or priority expectations in future user demand with high probability.

Setting up an information utility is one thing, making it work is a different matter. Where there is no question that the storage of indigenous cultural information is essential to our country's future, the key to creating value exists fundamentally in the underlying currents of TBDF. In this day and age, we cannot afford to limit the flow of our cultural information strictly to a domestic audience. Our information must circulate internationally if it is to increase in value. The argument that the rest of the world will not pay for Canadian information simply must not be entertained. Given the proper thrust of distribution, the international audiences will likely become just as blatantly parasitic on Canadian information as we are currently on theirs. If we were moving across their screens with any regularity, they would begin to show signs of our influence. They would eventually learn how to project our fantasies.

I can tell you because I know: our international distribution is weak. Our outbound traffic is a dribble. The interactive channels are set up for two-way traffic, but they're jammed with incoming signals. To reverse this tide, we must develop strategies for strengthening our cultural industries. The first step is to process their data into our information. Then we must send back their data, disguised, and improved, in the form of our information, to their screens for consumption. The Japanese have done this value-added reversal beautifully with hardware. They have taken American innovation, manufactured it on their assembly lines, and sold it back to the American market. We can do the same with their software. We can take their television and turn it into our television. By transforming their programming into ours, we manage to escape from our chronic stasis of total cultural passivity. If we aggressively redirect a processed, reconstituted American programming back into American homes, we will instigate a further derangement of the American self-image. The way it currently stacks up, we literally thrive on their culture without doing them a bit of harm. In the near future, a culturally parasitic nation like Canada may or may not prove to be harmful to its host.

Transborder data flow is governed by two primary factors. The first factor is the pressure generated through modern dissemination techniques. If our cultural industries were producing a high-quality cultural commodity at a competitive price, and we were moving this information through adequate distribution channels, we might be able to create the presence necessary to participate in the international information exchange. The second factor is the draw of the foreign audience. My wife always reminds me of this side of the coin. Eleanor tells me the audience always calls the shots. She means the audience controls both the volume and the rate of the flow through the influence of their demands. In other words, a receptive or hungry market will increase both the volume and the rate of any information exchange. Don't misunderstand. We still have to prime the pump. We cannot expect the American audience to fantasize the Canadian cultural industry into existence. We have to devise the formal qualities of our own identifiably Canadian information, and we have to convince Americans through our cultural commodity that they are just like us.

You might ask why we would think of participating in these culturally aggressive manoeuvres against the people of the United States? We simply have no choice but to do so. Unless our scientists figure out a way of screening our huge country with an electromagnetic shield, we must accept the reality that we are doomed to succumb to a saturation of American data and information. We have no way of protecting ourselves. Will we ever reach a state of cultural sovereignty in our information environment? Some of our people will. I believe my children feel their own culture is above all others in importance. They are living proof that cultural sovereignty is a perceptual fantasy. Our children have the power of imagination necessary to form the constantly changing images of their future. Later in life, they will surely construct the world they imagine today. They will live for the creation of their own image. Naturally, they will contract healthy levels of self-respect through the realization of their fantasies.

All of this adds up to one thing for certain. We must learn to determine value under a different set of rules. Money will continue to talk, but we will always be able to stretch the truth farther. We must be prepared to identify and isolate our obsessions. When I am weak, it is because I have failed to pinpoint my obsessions. I am often weak. My mind wanders through the channels as I roam in and out, over and under, and around every side. I'm riding the undulated energy of an elaborate scanning program. When I'm at work, I am witness to an incredible range of information exchange. Often, when I find myself locked in the scanning process, I find my mental imaging is totally incoherent. When your mind is virtually full of noise, when you crash from overscanning, you tend to feel weak and insecure. Every day at work I struggle to maintain a self-conscious distance between my mental notes and the phosphorescent readout of that VDT.

As I've already explained, I monitor TBDF for the Canadian government. I measure TBDF every working day. TBDF is something you will have to imagine for yourself. Although it may develop clearly in some of your minds, it remains an incomprehensible display of motion for most of us who read the machines. We make a lot of mistakes. We get caught in the close-ups of the traffic, in the detailed characteristics of the flow. We have one hell of a time determining value. Sometimes I think the best way for us to understand is to look straight back over our heads into the clear, boundless sky. The electricity of a cool solar wind rips into the vastness of an open field of atmospheric blue. The value must lie in the truest sense of the light as it reaches us. I don't know. As we soar within the strength and beauty of our machines, we must all suffer through the fragmentation of our fantasies and the constant dissipation of our obsessions.

Here at the Department of Communications we measure TBDF. Through our efforts, we are making a serious attempt to ascertain a clear delineation of our electronic cultural heritage. No matter how much cultural data we assemble, the general delineation of our culture remains obscure. The best way for all of us to see is to look straight back over our heads into the clear, boundless sky.

This essay was originally published under the pseudonym Andrew Czeszak. Czeszak is Tom Sherman's mother's maiden name.

I Get Tired 2001

I admit I get tired of managing information all the time. It would be nice to just let go a little more. To go with the flow. I used to have a life after work. I was paid a salary to keep things organized in the office. When things were crazy busy, my job was to follow procedures to keep things moving and in tip-top shape. When things were quiet, my partners encouraged me to stir things up, then to put things back in order before I'd go home. That's what white-collar workers have always done: they use telephones and computers to manage information, to stir things up, and to put things back in order.

My life got more complicated when I got hooked up at home. I set up a desktop and started working in the evenings and on weekends. I found I could straighten out the messes I'd made at work in my free time at home. I could get a lot more done by extending the office into my time off. I resisted the laptop thing because I needed some time to think, and commuting was the only time I had for myself. It seems as if I've always carried a cell phone on the road, for peace of mind. I use it mainly to keep in touch with my partners and family. I seldom initiate any business when I'm on the road. I know this limitation goes against the grain. I do answer work-related calls and react when I have to. Mostly, I listen to music and read novels. I do like to talk with my mother when I'm commuting.

The other day I realized how sick I am of managing information. I feel like a damned cleric or a librarian. Most of my waking hours are spent opening and closing files, putting things back in order, hacking away at the snarl of traffic in my in-box, launching reassuring messages in response. Staying in touch with people I hardly know. I thought my life's work would be more creative. I thought I'd be the one making the messes — interesting, vital messes. Instead, I find I'm merely an extension of this year's software. I use the latest, most powerful software as a template for my highly efficient, productive behaviour. I'm consistent, methodical, and orderly. There are loose ends, but these loose ends dangle in the Never-Never Land beyond the control of the system, the network somewhere beyond my desktop. I think about these loose ends when I'm commuting, between the time spent lost in my music, or in my novels, or talking to my mother.

TOM SHERMAN

People Want to Know What Other People Are Thinking 1999

There are excellent business opportunities in the fields of new media and information technologies. The digital culture is evolving quickly, and new markets are emerging and developing at breathtaking speed.

I overheard a couple of experienced information producers talking about a number of projects they launched way back in the old days — way back in 1995. One of these IPs turned the conversation to the present, saying these are incredibly exciting times we're living in. I thought to myself we all have to play the cards we are dealt, and life is short and precious, and all times are exciting when you're living.

Things are moving very fast today, and staying nimble is the key to survival and growth. Whether you're an artist or a hack, there isn't enough time to put things on paper and expect to see them realized. By the time you write up a proposal and flog it to prospective backers, someone else has acted on your ideas. Nothing succeeds like success. Money is quick. It takes money to make money. There's a lot of money migrating to the Net. Everything is migrating to the Net. Music, TV, movies, art, real estate — you name it — content is being transformed into new media in waves. Digital tidal waves. Investing money in new media is easy. Getting something back is another matter.

The first question is how do you leverage eyeballs to your Web site? Once you have them, how do you transfix them? How do you immobilize them and shake them down? How do you put their asses in your seats? Can you translate this attention and temporary paralysis into commerce? At what point is it safe to try to move them? Can you deliver objects they can touch and feel, things they want, or even people they want to meet in the flesh?

You can't make any money with advertising alone. Advertising alone is dead on the Net. In 1999, the average cost of advertising on the Web is US$52 per thousand pairs of eyeballs. You have to attract a lot of attention to make any money with advertising, and you can't sustain it. The way to go is advertising plus e-commerce on the back end. Get them for looking and send them home with something for their trouble. You have to have goods or services to sell from your site. Today, e-commerce is mostly click and ship, but shipping will gradually be eliminated by direct download. Boxes versus bandwidth. Of course, there are always going to be things you can't download. Pictures of naked bodies are simply ideal for direct download, but prostitutes still have to be shipped.

As you can imagine, setting up a venture in new media is risky business. But there are ways to cope with the hazards. Management is a science, after all. Operations research and organizational theory, applied successfully for decades in traditional manufacturing, are now applied effectively in the management of new media companies. Ongoing operations research provides algorithms for handling difficult decision making, decisions that have to be made on the fly within high levels of uncertainty. Organization theory now includes a range of horizontal or nonhierarchical management strategies. Employees can be given a sense of freedom and independence, but they can be exploited in much the same way new media companies exploit their customers: by systematically tracking behaviour and anticipating and facilitating the delivery or extraction of relevant information. People want to know what other people are thinking.

Staying one step ahead of the competition and the customers demands considerable efforts in research and development. Horizontal, or more organic, management structures permit the integration of R & D into daily service and production routines. Teams that are emotionally healthy and intellectually stimulated develop new methodologies quickly and grow content naturally. New media companies manage creativity, and wherever creativity is encouraged, there will be cost overruns. Take programming, for example. Managing programmers is like managing cats. What you want to avoid is R & D on the left side of R. Good luck. Given free rein, creative people will always expand their mess. Even in the best scenario, creative types will always want to expand the number of ways they can do the same thing. Managing creative people is expensive. Containment is absolutely necessary. Containment is accomplished only through structure and overt surveillance. People want to know what other people are thinking. What you want is R & D on the right side of D. That's product. The development of something people can use. Something people will buy into.

The new media business rides the new culture of convenience. All network culture is about making connections and getting what you want. Immediately. Everything in the world is potentially available 24/7. Content is important, but content is useless unless it is easy to access and a pleasure to consume. The key player in new media is the usability architect. Successful companies approach every endeavour with a usability analysis first. They lean heavily on their usability architect from the get-go. Usability factors drive the initial development of a project description, the targeting of audiences, and the determination of stakeholders. Application structure and site, and interface design rationale, are not left until the very end, but determine and control the development of a project from its genesis.

Before the launch, essential issues like hosting, anticipated community interaction, and site maintenance must be dealt with. And, of course, what are your e-commerce options? Do you intend to update and expand your site? If so, what are you doing in terms of stickiness? Today's user has the attention span of a housefly. What percentage of the project budget is set aside for updating and maintenance? Information refreshment. Or perhaps you are just going for the quick hit, assuming early abandonment. The choice is yours. In these early stages of the information era, leaving abandoned sites on servers, with or without active addresses, is not a big deal. From an economic perspective, rotten information, digital trash, hasn't built up to a problematic level yet. Information garbage disposal and recycling are not yet profitable.

IPs must think about their work from the user's perspective. In the simplest terms, what is the project? After you have an explanation that will hold water, who are your lawyer and accountant? Financiers will always want guarantees when projects appear insecure. Most will want to hold material equity against virtual equity. Nothing succeeds like success. Diverse, well-grounded corporations and companies with material assets can move quickly in uncertain, ethereal territories. Ideas are ten a penny. Never put money into business ventures so far ahead of the competition that there is supposedly no competition. Sweat equity is a factor, but it's never enough to close a deal. Stay clear of public money. It is too slow. Professional development scenarios are for losers. Governments do provide market assistance; offer grants or loans to boost budgets; or distribute survival funding. Bureaucrats pretending they are in business make things crawl with their political instincts to cover their asses.

Demos, pilots, and trailers can be constructed and pitched without worrying about copyright. Sketches of ideas are made quick and dirty. Stealing or lifting are essential shortcuts when trying to get out of the gate quickly. When you are pitching, anything goes. You know damn well that copyright is theft, a legal structure that is set up to block and encumber newcomers. But once you have a toehold, once you have something that resembles a property, never give up your copyright. Launches will fail, projects will stall, but copyrights are copyrights. Always maintain your copyrights.

The new media business is a resource industry. Audiences are cultivated to be mined and processed. Sets of eyeballs are attracted and sorted like eggs. Eyeballs are lured into portals, or general intersections of the Web. Portals are like local neighbourhoods or watering holes. In the portals, the audiences are divided into special groups, into vortals. Vortals are vertical portals. A vortal might offer vegetarian recipes or a chat line for specific monkey business; say, sex between large consenting primates with missing limbs. Links and analogies may be made between the content of related vortals. Thus hortals, horizontal, lateral, or associative *-ortals*, may be formed, creating a buzz between vortals. The gap is always where it's at. The object is to grab people and hold them, without letting them know they are captured, permitting them to move around in a loop. Companies operating portals, vortals, and hortals form partnerships and affiliate relationships. If a portal sends a user to a vortal, then the portal expects to receive compensation from the vortal for the delivery, and vice versa, and so on and so on.

These people are the content aggregators. They mix content from diverse locales, making something whole out of bits of this and that. They bring together many parts, divergent forms and styles, into aggregate substantiation. They take sterile, minimal, empty landscapes and mix up a new world, making things more naturally complex and substantially whole. Aggregators make the world up from scratch, over and over again. They create value by refreshing or updating the mix — or from the user's perspective, the fix. They are in the business of refreshing reality.

Everybody's killer app, oddly enough, is still e-mail. Everyone seems to be investing in the English language. It's a deadly virus in its own right, a plague on other tongues. Other languages are abandoned as droves migrate into English, stretching it and murdering it with their own brutally pragmatic style. It absorbs everyone in the interest of global communication. English is the common ground for a new formalism. If you think about it, the English language itself becomes the content. It doesn't matter what people are writing. The language is all you can see after a while. The Web is even worse. It is a horribly stiff, awkward conglomeration of technically restricted design modules reducing creativity to insignificant twitches too often interpreted as societal glitches. English is a predominant medium like video or the Web. These media are ubiquitous and common and virtually free for all to invest in. It is such a consistent world, a world of homogenization or loss, of stifling commonality and evaporating difference — but also of gain, of profit, that certain sameness of a shared, familiar reality.

Sometimes what you can actually license defies logic. You can license anything. Just remember to retain your copyrights. The media will come and go, but content is constantly reborn, like a sponge in water. CD-ROMs, all the rage just a few years ago, are dead, but the news hasn't reached the K through Grade 5 publishing market. Kids' parents have to buy something. Come to think of it, DVD-ROMs are actualizing the promise of CD-ROMs. With everything being downloaded, videotape is dead, but it is still cost effective. Special effects are hardly special anymore. A $40 million, mid-nineties computer-animated movie can now be made for $6 million, maybe less.

The key to new media is interactivity, and the basis of interactivity is cause and effect. Physics governs behaviour. Click and look, click and modify. Register your vote. Express yourself by selecting what you want from a field of available options. The real e-commerce is data mining, or surveying and developing a user-base as an information resource. Exploiting the potential of selling your users' preferences and behaviour to advertisers, marketing agencies, law enforcement officials, and parents. People want to know what other people are thinking.

Our Web-based radio station and record store produces a stream of user profiles in real time. Our users are locked in and interacting, while our data-mining clients are organizing new products and services, for all practical purposes, in real time. Our data-mining operations currently focus on twenty-four-year-olds. That's the vein of gold in demographics. Twenty-four-year-olds are the trend-setters. We target Web-literate audiences with mockumentaries or other forms of fact-and-fiction hybrids. Information as entertainment. The news set to music, serious music. The major thrust of our content is the constantly updated fantasy. With the kiddie market, this focus is easy. Dinosaurs that cohabit in a friendly manner with mammals, like us, are very popular.

The new media business continues to show downward trends in demographic targeting. The ultimate audience segment is the seven-year-old. The industry's goal is to seduce, manipulate, and exploit young children and by extension their parents. There are pedophiliac tendencies in the new media business sector. They are aiming their projects, affectionately called their *babies*, at the kiddie market. But they speak of noble intentions, positioning games beyond the realms of bloody carnage, or wanting to author games for girls and grown-ups and even seniors. As an immediate goal, they want to open up the stock-market for preteens.

Emerging markets on the Net will replace traditional markets. Markets decide what gets done. Audience preferences or user profiles will eventually be sold back to the users themselves. Audiences will pay for information about themselves. People need to be told what they desire, what they need. After all, they want everything but have limited resources. They need to know whom they resemble and whom they will likely get along with. They will need to know that they are valuable in their own way. I'm exploitable; I'm exploited; therefore, I am.

There is a whole industry based on dressing up misshapen, unappealing, lifeless content. We're going to make the navigation elegant and the site easy to use. We're going to make it look hip, irreverent, edgy, cutting edge. When we get done, your site will appeal to twenty-four-year-olds.

Ultimately, we're talking about utility for consumers. Transforming and migrating content, from film and video and every other medium under the sun, to the Web, is not really the primary focus of our activity. The main objective is creating utility for consumers and then tracking their behaviour while they are enjoying this utility. The user is the content of any new medium.

Is the nonlinear, interactive architecture underpinning the content the essential substance of the Web? In a culture of convenience, where navigation and utility cast a shadow on content labelled as such, the *content* is probably nothing more than a lame excuse for enjoying navigational freedom and elegant, satisfying utility. To make an analogy, the sports utility vehicle is more enjoyable than the landscape it potentially violates.

Navigation is not just the utility of moving users efficiently and effectively through content; it follows conventions and carries cultural associations as a process. Aesthetics are the glue between form, process, and content. Nuances and subtle twists and turns accumulate into real substance. Analogies and metaphors are constantly being created in the process of information exchange. Why is the discussion of aesthetics taboo when parties are trying to raise money to back a new media project? When will aesthetic issues be openly discussed by the new media business community? After all, people want to know what other people are thinking, about this, and that, and the other thing.

TOM SHERMAN

Millennial Spurn 1999

A Circus of Compliance

I know there are excellent business opportunities in new media, but what other kinds of opportunities are there? I'm not interested in business.

Everybody says new markets are emerging and developing at breathtaking speed in the digital age. Well, I don't want to move more quickly. I like to move at my own pace, the way I feel comfortable working. As far as I'm concerned, mobility and speed add up to nothing but new levels of superficiality. Timely superficiality is still superficiality.

All this hype about how we're living in exciting times is just another way of saying there's absolutely no sense of history today.

People say that things are moving so fast that being and staying nimble is the key to survival and growth. Being nimble means one is lively and agile. Being nimble is a mindset. The nimble must also be realistic and objective when confronting changes in the environment. The nimble are willing and capable of making adjustments. They are pliable and malleable.

The future will be run by compliant contortionists.

> > >

The New Media Are the Fried Chicken of the Information Age

People say things are moving so fast that it's a waste of time to write up proposals, because other people will just move on things, and you'll be left behind in the dust. This reasoning seems to be a very good argument for doing nothing or everything immediately. We seem to be witnessing the "maturation" of a volatile, half-baked culture. If everything is so rushed, it'll add up to something like a fast-food culture. Industrial-strength, dim-witted gestures will prevail. The junk will pile up. This junk culture is mounting at an accelerating pace right now, day after day after day.

Everything seems to be in translation. Nobody seems satisfied with original material. Everything is a twisted version of something else. Content, like chicken, is being transformed into new media, the fried chicken of the information age. Information is being processed and served on assembly lines. Industrial centres for new media production and distribution have been miniaturized to desktops and even laptops. Individuals are organizations, solo incorporations.

If you are reluctant to invest in such a culture, invest in something else, like flowering trees or canola oil.

> > >

What We Lose in Depth We Gain in Range

The culture of convenience is superficial, more than anything else, so why try to hide the fact?

We are moving toward a world where everything is available 24/7, whatever the long-range consequences. Do people really want to go faster and farther at all cost? We will find out soon enough, because nothing grows very quickly, or spreads very far, unless there is a time and space for it.

We're heading toward a paper-thin culture, but what we will lose in depth, we will gain in range. Although culture today may seem instantaneous on the surface, behind the scenes there is planning, planning, planning, and more planning. Today's superficial culture may look quick and dirty, but it is not. We've got to automate some of this planning to get the costs down.

Automating culture requires scientific management techniques. Management today is a science, after all. (Some feel this progress toward scientific management is society's greatest problem.) We can deny it, but more and more management is taking on the ideology and methodology of science.

Business today is not about keeping your word or building trust. Today, businesses are run on math. Once, the bottom line was the almighty dollar. Now, it is the almighty algorithm.

> > >

Management as Historical Agency

Managerial responses to present-day reality are always problematic, because we have nothing but old techniques to apply in completely new circumstances. This complaint is the main beef of youth: old ways of doing things are applied inappropriately to their realities. Paradoxically, management science tries to maximize profits by applying new techniques to old problems. The past and future, memory and hope, are defined to a great extent by the inadequate responses of management science, resulting in the aggravated frustration of youth. This outlook may yield some useful insights in terms of anticipating, controlling, and subverting currently unmanageable emotional behaviour in the future.

We always fail when we apply old management techniques in new circumstances. We can choose to see this failure as an unavoidably flawed but logical, pragmatic attempt to solve today's problems, or as a simple exercise designed to moderate the uncertainty of the future by historicizing the present.

> > >

Acceptable Levels of Raw Personal Material

Even though we live in a junk-food culture, where life is squeezed out of the product before it reaches the consumer, significant levels of raw personal content will permeate the market as the tools and means of production and distribution continue to be decentralized and become ubiquitous. Information providers work at home. The information factories are miniaturized as desktops and laptops. Such a vastly decentralized production and distribution system cannot be regulated easily. Raw personal material (RPM) is fast and vibrant and abundant. Raw personal material will continue to seep into the streams and torrents of junk culture.

What's surprising to me is how intolerant most people are when it comes to accepting raw personal material in their cultures. Even though computers have completely eliminated the differences between ordinary telephone lines and broadcast channels, people are still reluctant to embrace real, ordinary humans in their entertainment and art. They treat raw personal material like unwanted insects in a perfect nature. Most people are disgusted or horrified by flies, mosquitoes, and ticks in the garden.

Closer to home, it's like finding insect parts in a jar of peanut butter. Government regulatory bodies determine the acceptable or tolerable levels of insect parts in peanut butter. Basically, the insect parts have to be imperceptibly small. It is impossible to produce a jar of peanut butter without including a certain amount of raw insect protein, parts of legs, wings, bodies, and so on, but if they are ground up fine, a small percentage of raw insect material in peanut butter is acceptable. It has to be.

It's like this in culture, too. It's impossible to produce culture, entertainment, or art that doesn't include raw personal material. Entertainment and art formalize, fictionalize, and dress up RPM. People still like the *idea* of a culture springing forth from the people themselves. Before cultures were dehumanized and rendered totally artificial, cultural trends used to emerge from the people. Today, raw personal material is usually categorized as interference or noise. Talk shows, reality programming, and listservs feature raw personal material groomed or organized for consumption. With the blurring of telephone lines and broadcast channels, governments seem content to let markets set the limits. Junk-culture markets dictate acceptable levels of raw personal material, whereas regulatory bodies set the limits in other, more stable sectors.

> > >

Quick and Dirty Perfection

Ideas are ten a penny. Actually, the reverse is true. There are damn few ideas that'll hold any water. But being light as a feather has its advantages. Economies and markets are like anchors. Conceptual art never had any real economy, and look how it took off. Network art has the same upside. Nobody gives a shit, but it is spreading like wildfire. It makes you think twice before you make something heavy to carry and difficult to store.

In this light, let's examine the drawing versus painting debate. If you ask people what's more important, they'll pick painting because colour is hard, painting techniques are more difficult to master, mistakes are often permanent, and so on. But let's face it: craft time is overrated, and material persistence and permanence are outdated concepts. In other words, why move beyond the sketch? If people have so little time to digest their culture, then why spend time making things that last?

So much of what artists create is the result of taking the time necessary to see or recognize something as it is, and taking the time to represent this something elegantly, or taking the time to tweak it until this something sings — the time spent on the work is somehow saved or "banked" in the work and presented as the basis of the work's value.

We learn to appreciate things over time as our experience builds or accrues in layers. With information refreshment, there will naturally be layering through multiple points of view.

It is one thing to make art in materials that last, creating the illusion of permanence. It is another thing to plan ongoing maintenance for a work of art, so that the work will be cared for and maintained in the future. This plan conjures up images of works of art that come complete with custodians or maintenance workers contracted to care for them.

Much conceptual work is a documentation of ongoing obsessions, taking time and compressing it to create value through documented procedures and processes.

Time-based art, improvised in real time, features the artists making the work in the same time it takes to experience the work. In a culture of convenience, 24/7 culture, it makes sense to produce information in the exact amount of time it takes to consume it.

When business-oriented people talk about demos, pilots, and sketches, they say that everything is fair game when you're putting together a pitch. If you need a soundtrack, use a track off your favourite CD. Steal what you need. Cut and paste. Move as fast as you can to make your point. The question I have to ask is why move beyond the sketch?

> > >

No Secrets

In the mature information economy, information will be exchanged, information-for-information, and those without information to trade will be dirt poor. As it is now, most people only have a tiny bit of personal information (raw personal material, RPM). It makes you wonder what will happen when they lose all their privacy.

People will get food stamps for telling others their deepest secrets — things like how they have been having sex with their mothers or their disconnected princess telephones.

Exchanges of information are economic transactions, plain and simple. The obsession with translating information into money, into financial currencies or commercial properties, is only meaningful in this period of transition, and it is overrated.

A transaction is a transaction. In a mature information economy, information is exchanged directly without being tethered to money.

Those without information will be poor, as is already the case, and those who lose their privacy will be the poorest. What will happen when people have no information and no privacy?

In the information economy, people are harvested like trees or minerals or fish.

Right now, people are being harvested like trees, but soon they will be cultivated, planted, and harvested like an agricultural product. Then, in turn, they'll be manufactured, enslaved to provide a service, and then turned into outlets for the accumulation and release of something called knowledge.

> > >

ACCORDION STRUCTURES (PORTS, VORTS, HORTS, -ORTS) OF AUGMENTED HUMAN DEVELOPMENT

There is a sexy sense of fragility in the electronic, digital culture. Coupling is necessary. Being and remaining aloof is too risky. This vulnerability is reflected in our tentative, shifting identities. It's a sketchy existence at best; such a shaky cognitive in-firm-a.

People are trying to charge magnetic sites, where eyes are drawn, sucked into the quicksand of attractive images. Portals are the sites of general communal interests: water, food, shelter, companionship. The vortals, vertical portals, are where social selves are transformed into individuals. The vortals are intimate spaces, the private rooms. Then in hortals, the reflattened vortals, the temporarily definitive selves are let out into a more spacious uncertainty, like range animals, encouraged to collect and cultivate their cohorts and cults, to reestablish societal integration, within concealed limits.

In other words, the human development scenario is now regulated and paced by technology. People alternate between introversion and extroversion, exercising their instincts when facing accelerated cycles of self-definition and self-disintegration. They reestablish, then reject, social ties, rediscovering the feeling of belonging, then escaping commitments, vanishing into thin air.

Restraint, not tight bondage, is in order. The gap is always where it's at, even if it's tethered at both ends. The objective is to grab people and hold them without letting them know they are captured, permitting them to move in a loop, struggling without too much discomfort. Appropriate or sensitive containment should provide all the trappings of being held. Give them the illusion of choice. They must think they can unplug or escape whenever they want, but they can't without terminating themselves. This entrapment is like living in television at the end of the day, holding the remote, scanning, roaming the channels; being completely exhausted, yet unable to hit the power button, to kill the television and by extension oneself, ending one's engagement with the "world."

Thus people experience the feeling of being restrained by a firm embrace of their consciousness, not a comforting embrace, but an embrace nevertheless. The somewhat comforting, uncomfortable sense of being held. Arrested. Industrial-strength belonging.

In vortals, people are rediscovering the feeling of belonging. Then before things get too chummy, they are redispersed, sent on their way, hortalling out into another void.

We value the vacuum or draw of empty space, as we move out into an area of low pressure, where our ideas are extracted, drawn like the wind, depending on the prevailing conditions or circumstances.

We are talking about the development, the manufacture of an essential frontier: reverse-engineering a private space.

Portals, vortals, hortals, and subsequent -ortals. We move through an expanding and contracting spatial labyrinth, an accordionlike sequential structure for harnessing human development. Ports, vorts, horts, and -orts.

We witness the oscillations of animal magnetism (attraction, confusion, boredom, rejection, repulsion) in these accordion structures of augmented human development.

> > >

Cultural Geology, Chemistry, and Biology

When I think about content aggregators, the fabricators of digital multimedia, I get the feeling they are making or forging new rocks, not cultural objects. The word *aggregate* is so closely associated with geology. Humans now exert geophysical forces, forging new aggregate compositions. This process isn't about sticking things together with a glue gun, but it is an act of fusion, the permanent bonding of previously discrete elements. And in terms of the geophysical, it is the elimination of geography, of specific location. It offers the view from the macroscope. The different kinds of rocks and their respective mountains are compressed, and locales, definable differences of place, are eliminated.

What is the magic process, the power that forges aggregation? Is it a massive, sustained pressure, like the power that crushes carbon into diamonds, or is it a sophistication of knowledge, simply knowing precisely how to do it?

So much of the electronic culture is paper thin, only skin deep, or dilute, like the weakest solutions. The music, TV, movies, whatever, all seem incredibly thin, adhering inadvertently to the 20 per cent rule (cultural elements today generally have only 20 per cent resolutions). Today's pop music, sitcoms, and movies are empty, formulaic time-expanses, the dilute cultural products of industrial artists going through the motions. These incomplete cultural facades are just the right consistency for mixing. Everything has to be nice and thin and dilute to be mixed together productively, successfully.

The predominant idea in digital multimedia is the creation of density and mass through aggregation, although in biological terms, genetic engineering permits the fusion of biological traits through a physically articulate, more precisely targeted process of aggregation. The idea of accelerating evolution or simply growing varieties of Frankenstein is far more sophisticated than multimedia's characteristic parallel sources and overload (content aggregation via density and mass).

> > >

Hybridity, Transparency, and Loss

Increasingly, there is nothing new to watch, or to listen to, or to think about. This loss is due to the homogenization or blend of cultures. It reminds me of the announcers on a Canadian Broadcasting Corporation radio program saying we no longer have to call world music "world music." Canada is a multicultural nation; therefore, world music produced and appreciated in Canada is "Canadian music."

Nationalism is the ultimate aggregator. Diverse cultures are amassed and totalled into composite national cultures.

> > >

The Cost of Killer Apps

People are exchanging messages in record numbers, writing and speaking in the language-of-the-gap. English is the filler between cultural differences and diversity. The machines speak English. Machine English is in its own way a kind of valley speak, because even the machine voices lift the end of their phrases, making statements that sound like questions, telegraphing that they are not through talking yet, they have more to say. As with all in-between languages, there is a radical, breathless evacuation of meaning.

English is the linguistic commons, like the sea, or a tract of land jointly owned by neighbouring parties. Nobody really owns it, and thus nobody cares for it or protects it. When speaking the language of the gap, of the commons, whether it is English, or video pouring from a digital spigot, or the rigid, universal design protocol of the Web, there is an unending sameness. Anyone refusing to conform, acting in an anomalous, gritty, or jagged manner, is seen as an anarchist, as the incomprehensible enemy. Creativity is also viewed this way by conservative, traditional minds.

It's the Monoculture with a capital M versus cultural diversity, the social descriptor for biodiversity. It's suffocating. Didn't there used to be more noise? The all-pervasive English and ubiquitous video and endlessly uniform Web design are oh-so-common and virtually free for everyone to invest in. Except with digital telecom, you have to be able to afford access, and you are best served by driving a big, fast machine, a G4, or G5, or G6 — whatever is faster and cheaper, cheaper per memory unit. Demands for speed and scale and range increasingly dictate more and more overall expense: faster and cheaper, except when there are elaborate architectures of thought to convey; when the ideas exchanged are more and more complex and elaborate in scale; ideas best conveyed in a multidimensional display.

To remain optimistic, we must never underestimate the human imagination, even though we seem to be at the end of our tethers as fewer and fewer people seem to be able to imagine anything other than the models or templates for creativity and invention they have been provided with or force fed, by Disney or Universal or whomever. Industrially produced architectures of thought generate imaginative uniformity. Universal architectures of thought are generated by universally linked or shared minds. Common architectures of thought are valuable in that they create order and allow coherent, complex thoughts to be communicated, transported from like mind to like mind. Stereotypes, after all, are useful knowledge structures.

> > >

Super-morphing Saga

New technologies will come and go, but human nature advances or regresses slowly. We always look to new technologies to stimulate social interaction. We want to buy friendship, dialogue, community, presence, influence, and ultimately respect.

We are the content of any new medium. First, the content is the people; then the medium itself; and, finally, people emerge again as the content of any new, but aging, medium.

It's a story of people transcending or overcoming media. It is a kind of supermorphing saga where people transform into and emerge from inanimate objects.

We are just as genetically hardwired as the birds and bees, with our instincts, our most fundamental behaviours, plus we have our religious conditioning, family values, societal engineering, class status, and politics.

We carry large complex architectures of thought with us wherever we go: architectures of thought unique in detail, but surprisingly uniform in architecture and overall assembly.

> > >

T Not P

Increasingly, we are bored stiff. We may be leading longer lives, but our lives are increasingly empty. They have a monotonous, flaccid tone. We are tired, made weary of constantly adapting to change. We are constantly asked to accommodate technological change, apparently arbitrary change. This flatness or deadness of heart is unnerving. Don't check for my pulse. I just want to be a dial tone.

> > >

Special Effects

Special effects are no longer special. They are expected, as the manufactured frontier of the explicit-unreal replaces the real and ordinary domain of the physical. With computing power becoming faster and cheaper, there will be even more computer-generated films with more and more elaborate special effects. And these special effects will become less and less special as we, in our physical domain, become more real and ordinary, and less special, every day.

The expansion of the human experience will be the materialized imagination. By firming up a mould or template for the imagination, we will guarantee the total collective uniformity of fantasy.

> > >

Interactivity and Vectors

The key to new media is interactivity, and the basis of interactivity is cause and effect. Being able to make a difference, even the tiniest difference, is what people want. They will accept, even relish, being a tiny speck of difference in a world of gigantic, all-pervasive sameness.

When we step back from direct person-to-person experience, our effect on the world is reduced in scale. It is as if we get excited about being insects again — being able to inflict mosquito bites — but we would be so much more impressive if we could deliver West Nile fever or malaria. I want to be a vector, so when I make my little dent, or bite, it has big, magnificent consequences.

We like interactive media experiences. We like to be able to make a little dent in our reality. Interacting with the world helps us form and reform our identity. Some are satisfied constructing their identity with cheap kits, like a handful of choices we can click in a menu, or the inane questions of a public opinion survey. Remember Mr. and Mrs. Potato Head?

Cheap kits don't help us to refine our identities. We need to be recognized, documented, insulted, assaulted, and attacked to sense our presence in the world.

There is nothing worse than facing indifference head-on and realizing you are being ignored; that you are invisible, totally inconsequential. In the information age, the struggle is between difference and indifference.

Hence the deadly little bites of terrorism. All little bites that make a difference are problematic in the state of indifference, whether they maim or kill or gum up the works or simply manifest in unorthodox behaviour.

> > >

Bored People Are Dangerous

People want to be mind readers. They're nosey. They want to invade other people's consciousness. They want to control mindshare. Ultimately, they want to remote-control people. Exploitation is constantly being redefined by advancing technology.

On the question of how to leverage eyeballs to your Web site, what are we looking at? We're looking at a mob of exhibitionists vying for your attention, constantly upping their levels of exposure. They're playing chicken with the real probability of psychological implosion.

Plucking eyeballs out of the skulls of the on-line masses is disgusting, by the way. This obsession with attention is sick.

Once they have the attention they so desperately seek, they'll turn aggressive. They'll flip from desperate exhibitionists into control freaks. First, they'll author media releases that function like hypnosis. They'll cast a spell. Then they'll fashion media packages that pack the wallop of a stun gun.

Control is based on predictability. Surveillance and reducing uncertainty is at the heart of the matter. Eliminating risk and theft and loss is the main obsession. People want to know what's going on when they have something at stake. Otherwise, they couldn't care less. Unless they are bored stiff. Then they'll watch anything. Bored people, given the opportunity, will just bug people at random to find out what they are doing.

> > >

Bulk Erasing and Repurposing

People say that information garbage disposal and recycling are not yet profitable. But I've seen forecasts of massive recycling campaigns generated by the people for the people, where everyone recycles cultural garbage for their own amusement. They will make things out of abandoned software the way people used to make floormats out of bottle caps, or decorative chains of dead novelty watches from fast-food outlets. And then there is the information-demolition trend. Blowing up or otherwise disintegrating rotten information is very entertaining, and it is a service our societies obviously need.

> > >

History Is Spatial

In China, you've got free enterprise zones, totally wired zones where pan-capitalism is practised in support of the larger communist project. In the same country you've got fifteenth-century farming and twenty-first-century electronic commerce. History is spatial.

In North America, you've got people heating with woodstoves and riding only bicycles, and some of them are still living with single, fixed telephones or no phone at all. And then you've got people obsessed with digital communications, living with wireless phones, e-mail, satellite dishes — you name it. Sometimes this contrast is still the difference between urban and rural realities, but it is more fundamentally a decision people make to define what kind of life they will lead. History is an elective option. People define their lives by how many layers of technology they have between them and their neighbours. Of course, these decisions are often income-driven and have status written all over them. People struggling to make a living don't have the privilege of being able to determine their teledensity. In this technological age, there are going to be anachronisms in your neighbourhood. History is spatial, but you can drive the time machine if you have the means to do so.

So there are those with the options and those not having the privilege of choice. And there is imprisonment inside these technological systems and in an exclusion from these systems. There is the difference between being able to make a difference and choosing or inheriting indifference.

In the developed nations, a number of people get to choose these differences. But in the developing world, where teledensity is low and the infrastructural gap makes being wired expensive or impossible, it is not a matter of choice.

Different levels of teledensity, whether chosen or not, define the time, the relative history, we live in.

Today, in the developed world, teledensity is sometimes determined by urban density and infrastructure but more often microsocially by human initiative and choice. People distinguish themselves from one another by adopting or rejecting technological time and space, the new territory. Choosing to adopt technology is associated with progressive open-mindedness (get with the program! jump on the digital bandwagon!), the utopian power of positive thinking. If people choose to hold back, to reject technological progressions, they are labelled Luddites, traditionalists, and conservatives. This liberal and conservative labelling is in fact misleading, given what used to be understood as the left and right. Being critical of or unwilling to adopt the latest technologies is generally associated with the traditional political left. In fact, there are high levels of ultra-right-wing-conservatives among the

digital literati. The evil dig the power surge as much, if not more, than anyone else. We see a lot of people using the new technologies to preach hate and maintain an unjust status quo. This demonstrates that life in the information age is not black and white, and leads to the argument that technology is neutral, followed by technology is always linked to military concerns, technology is gender imbalanced, and so on, and so on.

> > >

Parents, Police, Advertisers, and Marketing

Parents, police, advertisers, and marketing agencies are strange bedfellows, but they are clearly the purveyors of order in pan-capitalistic societies. Parents and police are countersupported by advertisers and marketing agencies, as designers and demographers work hand in hand to stimulate and maintain consumer demand and economic activity. Keeping people busy shopping and consuming, chasing their dreams and desires in work and play, supplants the enforcement of law and order.

> > >

News Music

I've been thinking about delivering a newscast set to music, serious music. Right now, music is used to introduce the news, or to segue between the news and the ads, or vice versa. Alarming events like giant earthquakes, hurricanes, transportation disasters, or military campaigns have their own theme music, usually an elevation of horns anchored with dramatic pounding drums. But such conventional news-music isn't mixed into or run under the delivery of the news because it would emphasize the show-business side of the news. Music would undermine the authority and weaken the integrity of the plain and simple truth represented by the nakedness of the newscaster's voice. The news could use some music, to move it along and make it more like a movie or a song. I know it's difficult to imagine the news delivered with authority in a musical voice. But I think it's going to happen. It's just a matter of time. Serious news-music just hasn't been written yet.

> > >

Intelligent Shopper-Agents

People say there is only room for so many markets. If digital telecom spells the end of geography, and market congestion and crowding isn't the problem — in other words, if space and traffic are not significant factors in market failure — then there are simply not enough shoppers. The Western economy suffers from a serious shortage of shoppers. That's why the ballooning middle classes in India and China are so tantalizing. Look at the numbers. Shoppers everywhere are looking for the same things. We're all hungry for identity props. We're shopping to shore up our identities. Opinion polls and consumer behaviour surveys abound. People crave feedback that will tell them who they are and what they are supposed to be interested in buying. People are like kids with Christmas catalogues: they want everything but have limited resources. They have to make intelligent decisions. Artificial intelligence research, and specifically the development of intelligent shopper-agents, may be the way to generate larger numbers of shoppers quickly. Intelligent shopper-agents, computer automata that shop 24/7 and never tire, could mushroom into a vast army of supershoppers, a formidable wave of aggressive surrogate consumers. Initially, these intelligent shopper-agents could fill and buoy failing markets. Eventually, such artificially intelligent shopper-agents could augment human shoppers in the West, permitting outnumbered Western shoppers to compete successfully for goods and services against the surging demands of the Indian and Chinese.

> > >

24/7 Culture

The new media business rides the new culture of convenience. All network culture is about making connections and getting what you want. Immediately. Everything in the world is potentially available 24/7. Content is important, but content is useless unless it is easy to access and a pleasure to consume.

Our data-mining operations currently focus on twenty-four-year-olds. That's the vein of gold in terms of demographics. Twenty-four-year-olds are the trend-setters. The new media business continues to show downward trends in demographic targeting. The ultimate audience segment is the seven-year-old. The industry's goal is to seduce, manipulate, and exploit young children, and by extension their parents, grandparents, aunts, and uncles. The demographic pressure points of the culture of convenience are 24/7.

We are in the process of forging a continuous, lifelong, twenty-four-year-old identity or sensibility. All kids will want to grow up to be twenty-four, until they've made it, and then they'll want to be twenty-four forever. Twenty-four-year-olds still have flat stomachs. They are the ones behind the proliferation of abdominal muscle displays, the bare bellies and midriffs. Abs are us! Part of the twenty-four-year-old thing is the pure attractiveness of young adult flesh.

We all need to know that we are valuable in our own way. There must be human values beyond income and wealth. If you want my attention, appeal to my vanity, address my needs for security, status, and companionship.

In the future, everyone will think they are the same age. Everyone will be twenty-four.

Before that, to meet the kids head on, the IPs call their new media projects their *babies*, and by making their projects their babies, their juvenile agents, they send their babies out to make friends with and to seduce other children. Seven-year-olds, having only looked outside themselves, have no understanding of the strange feelings they have inside. They represent total emotional vulnerability.

Dinosaurs are very popular as friends for kids. Even monsters big enough to eat us are our friends. What a relief! The equivalent adult content would be a situation where major multinational corporations are shown to be our friends. They could eat us or squash us, absorb us, or crush us, but they choose to be our friends. How comforting. They could literally kill us, but some of them will even help us get rich.

This era marks the end of the idea of savings; no more security for the spendthrifts. Now it's gambling and gaming, the lotteries, and video bandits, and playing the stock market with reckless abandon. Going for broke.

They are now teaching market speculation in grade school and junior high. High finance is a form of play. Start them young. After all, kids have no fear. They play the market with cold blood.

> > >

INFO-FEEDERS

We need to develop bird feeder-like devices and things like duck decoys for humans. We have to learn how to make electronic- or media-bird feeders, info-feeders, so that we may lure people into our neighbourhoods and by feeding them information, entice them to stick around. Images of flat stomachs with abdominal six-packs are excellent decoys for attracting twenty-four-year-old wannabes.

When people ask us what are we going to do in terms of stickiness, I tell them we intend to create a kind of cultural flypaper. Or for larger audience blocks, we've invented a new mousetrap. We are also developing digital decoys and duck blind-like cover, and for hostile audiences, we are experimenting with cyber-booby traps.

Info-feeders and all kinds of other info-devices are prone to breaking free from the world, and so info-devices can't really help connect us with nature. You know, the Land Rover is confined to the highway most of the time. Things always seem to slip farther and farther away as we get closer and closer to the perfect or absolute unreal.

Remember, if it looks like information, it probably isn't.

The Remote Volunteer 1997

Tom Sherman participated in Kunstradio's "Recycling the Future" intervention at *Documenta X* in Kassel, Germany, in July 1997. Kunstradio was part of *Hybrid WorkSpace @ Documenta X*, an information-based alternative (conceived and curated by Geert Lovink and Pit Schultz) to the conventional art exposition. Sherman, calling himself the "remote volunteer," sent e-mail messages to the *Hybrid WorkSpace* and eventually participated in a one-hour voice-music performance in Kassel via telephone from Port Mouton, Nova Scotia, Canada. In the following three e-mail messages, Sherman discusses the new information economy and speculates on the changing nature of work at the century's end.

> > >

As the Recycling the Future Team Descends on the Hybrid WorkSpace

July 16, 1997

As the Recycling the Future team descends on the Hybrid WorkSpace, I can't help but remember the future, as I once knew it years ago. We were always taught that the future would bring us the leisure society, a societal condition where we would be working hard every day to amuse ourselves, not to create material goods. Now I don't want to imply that the media artists and information producers setting up camp at the Hybrid WorkSpace are just working for themselves, to keep themselves busy, because of course they are going to have an audience besides themselves and each other. I shouldn't say they, but rather we, as I'm also revisiting a future where it will be possible to work in Kassel from my home in rural Nova Scotia, where I'm writing this morning. We will have the art audience and we'll have the broader potential audience. The art audience is certainly part of the leisure society we heard would develop by the end of this century.

To double back for just a moment, everyone, even artists and information producers of every stripe, knows that the global economy (the ecology of exchanges measured by symbolic currencies) is now driven by information and knowledge. We still have the resource economy (fish and gold and silica) and manufacturing (computers and office furniture and paintings) and the service sector (hairstyling, fish and chicken cooked with different sauces, banks), but the economic sector that is kicking ass is the information and knowledge sector. I & K. The I & K phenomenon is basically an endless description of how we can better manage the resource, manufacturing, and service industries. And looking ahead, how we can manage the leisure and tourism and finally outer-space industries. We now generate texts on how to manage our resources; how to be more efficient in manufacturing products (how to raise fatter chickens on less feed); and how to provide services that people want and need (how to serve tastier chickens at lower costs). The I & K sector is the how-to sector of the economy. The I & K explosion is a great time to be an artist or a consultant or any other form of scribe (decipherer, describer, cyberdescriber) because I & K workers make a living deciphering and describing things. They make "texts," not tables and chairs or sculpture or paintings. Images, sounds, ideas, concepts — and strings of same.

So the Recycling the Future team, and every other team setting up its data-processing or information-production facilities there (here) in the Hybrid WorkSpace, is getting ready to describe something. We're getting ready to describe the resource, manufacturing, and service sectors, and, of course, there's the leisure and tourism sector. But before we get back to killing time and amusing ourselves, let's examine the changing nature of work itself and how it provides us with the raw material of our identities, our psychological anchors or security blankets. The Hybrid WorkSpace is a Virtual Employment Zone. A VEZ: it's an "office" or an information marketplace, or a simply wonderful place where unemployed or underemployed people find the work they want to do. It's a work zone, where pure work, relatively uncompromised work, can be found or invented or taken up or on. It's not like a school, where societal roles are worn like business suits until they fit more or less and are modelled down a fashion ramp right into the world. No, the Hybrid WorkSpace is a more difficult, mysterious, ill-defined zone. The timeline for success is compressed (careers are not launched in a VEZ), and there's simply more media than there is action. In fact, there's no action there to start with — it all has to be created. So the Recycling the Future team, and every other team, has brought the tools for taking and making action and generating images and voices and generating a local, virtual explosion of description, a kit for making a text-bomb. And by text I mean virtual chickens and cows and hairstyles and close-ups of the horrors of the unemployed, homeless, and murdered, not sculpture and paintings or office furniture.

Now I'm imagining this Hybrid WorkSpace as being something between an office, a café, and a home. You know, the places where work occurs today, in the future (July 1997). Of course, today's work also takes place in cars and on bicycles and on skateboards (and on planes and trains) and while people are walking and talking. Talking is important, except when you're stalking. Information gathering is often a solitary, foraging activity. Back to motion, work today is often mobile and frequently taking place while you're in transition. Portable. If work can be anything that people want or have to do — and description is hard work and must be motivated by desire — then it can take place anywhere. Before I go any further with this somewhat staggering overview (please don't make me walk a straight line), it strikes me that this is a make-work project. You know, the government pays people to apply themselves in socially constructive ways to solve the problem of social volatility and instability. You know, it used to be just youth rolling around, drinking and drugging and fucking and "wasting their lives." Shit, they aren't describing anything, except how to party. Hell, there's still too much human waste in the youth sector — but even more distressing is the empty thrashing around of the middle aged. We have all these life skills and still we're bored and disenchanted and so critical and destructive. We need hard problems to solve, difficult employment, Virtual Employment Zones (VEZs) where resources are provided, some travel permitted, networking tools at our fingertips, but there's nothing else, except an inappropriate setting, a big art fair

(*Documenta X*), and a stream of leisure-class tourists, perhaps impervious to "real work." The Hybrid WorkSpace is such an extremely difficult Virtual Employment Zone. It's a junction where intriguing connections will have to be made, or the leisure class and its propensity for boredom will prevail.

It isn't just the youth and the middle aged who are wasting their lives, not describing anything productive; it's the elderly we've got to worry about. The governments will have to set up make-work projects like the Hybrid WorkSpace for the elderly as well. Shit, if I was seventy-seven years old, I'd probably have to lie about my age to get you to post my messages as part of your Recycling fiasco. Fuck, it's not just a make-work problem; having to be socially productive or constructive is like a prison for the elderly, who just want to cut loose after leading practically endless productive lives and then being rejected by the assholes now running the economy. If I were seventy-seven and I was generating Internet traffic and setting up so many Web sites, I couldn't remember where the servers were, let alone the point of my abandoned pages. Hell, I get e-mail from people I don't remember writing to now. The future is memory disorders. I just turn on my e-mail in the morning and weed through the in-basket. Who the hell are these people anyway?! I comb through the messages, trying to remember anything about any of them and then I make some mental notes on the work I'd like to do today, but then I just forget to move on these matters. So I make long lists, and they just keep getting longer and longer, on little stacks of paper and index cards. And then finally I send off four or five messages like messages in a bottle, and three or four of them never seem to have any effect, because I'm seventy-seven and no one gives a shit about my descriptions. I'm hoping they, the government, will set up a Virtual Employment Zone for people like me. You know, a real, physical place where I can meet people and trade things that I get off on and find my voice again. I know ways to party, too. I'm knee-deep in my own kind of human waste. Give me a chance, and I'll throw some shit at the tourists. Hybrid WorkSpace, my ass!

> > >

The Recycling the Future Team Has Unpacked the Equipment and They're Ready to Roll... Errr, They're Already Rolling!

July 19, 1997

In the future, people will work in the comfort of their own homes. God, it's lonely over here. People are so busy there that they haven't had a chance to assign me anything to do. I'm ready for my marching orders, but I haven't received any e-mail or faxes or phone calls or telepathy or anything. Don't get me wrong. I like working in the Hybrid WorkSpace. It gives me a sense of belonging to an organization and when you're working by yourself and for yourself most of the time, it's good to find yourself in an office with coworkers and colleagues and people who are just plain sharing the mission. I mean, naturally, there are problems whenever you work with people and depend on other people to get things done, but you just can't beat that corporate feeling when you believe in the people you choose to work with — and share the goals of the mission. Let's all join in and accomplish something! Anything!

Do you mind if I play a CD? I'm listening to "Sluts'n'Strings & 909," and I've got it turned up so fucking loud the putty is falling out of my windows. I swear to God, the big black shiny spiders in my cellar run and hide when I turn the Sluts on loud. That's one of the things I like about working at home. I can play my music as loud as I want without bugging my coworkers. Where I would probably be annoying as the next person there, in person.

It's rare for artists to find a way to work together, to agree on anything. And wasn't it clever to set up a physical site for this collaboration to happen. A showplace for interdisciplinary cooperation — or at least the pleasure of "playing" in parallel. You know, the independence and diversity thing. Autonomy and all that. The only thing I've ever heard of that resembles the Hybrid WorkSpace was a thing that happened a long, long time ago here in Canada, something called the artist-run centre. Back in early 1970s, the government funded these physical spaces where people could get together and meet, and the people put together media facilities in these sites, and they set about solving the problem of finding ways of connecting themselves to the world. You see, the artists and writers and poets and musicians felt alienated, and they wanted to change the world for the better. They hooked up with social activists who had some organizing skills and some knack for being annoying and provocative and they started to cook, making beautiful noise all across Canada.

The governments thought these artist-run centres were a good cause, plus they could kill two birds with one stone. They managed to stabilize and control and tame a bunch of, from their point of view, pretty kooky people, and they were able to make examples out of the projects that emerged. They showcased a group of citizens trying to make a difference in a sea of apathy. For good and bad, political correctness was refined in these centres. There had to be rules to keep people from killing each other. Collaboration can be nasty business. But they worked things out and promised themselves they'd continue to do good deeds for the taxpayers, in exchange for grants. That's what public money and such a bright spotlight can do for chaos: it can engineer consensual behaviour favourable to the taxpayer, for the good of society. Talk about the mysteries of emerging social order.

Most of these artist-run centres are deader than doornails today, after a quarter of century of compromises. But this story is history and we're supposed to be recycling the future, not mucking around in the past. Things happen very quickly today, and things will probably come together even more quickly in the future. I wouldn't be surprised if the Hybrid WorkSpace plays out the whole history and evolution of the artist-run centre movement in a couple of months. Concrete deadlines and real timelines can make people real efficient. This is good because government officials suffer from attention deficit disorder, just like everyone else. So we'll probably see a creative social movement compressed into a snappy model event we can hold up as an example of pure energy, intelligence, economy, and goodwill in the future. Maybe it will function as a catalyst for generating change or engineering employment or involvement or commitment in a world immobilized by depression and cynicism. I don't know. I'm rolling up my sleeves and I'm ready to pitch in, but I need something real, the real vibe from the centre. It's lonely working at home. Please send word of what I can do so I can feel part of the Hybrid WorkSpace in the immediate future, like maybe later today?

> > >

Recycling the Future: Coping with the Sameness of Continuous Change

July 23, 1997

I've finally heard from headquarters. I've received several messages, telling me to just relax about my role, that my contributions are valued, and that my voice is coming through loud and clear. It seems that's what they want: they want me to be committed to my work, to do what I feel is important for me to do: and to do it to the best of my ability. Apparently, that's what's happening there, and thus by implication, here. People are sort of celebrating their engagement with their work, the things they are driven to do, enjoying each other's company under the various degrees of pressure assumed as part of their self-imposed, self-determined workload.

It is as if all the people in the Hybrid WorkSpace, the people on-site and the many more distributed all over the world, have volunteered to toil together, almost for the pure joy of collective involvement. No, *engagement* is a better word to describe the commitment. The Recycling team describes a zone of mutuality, where individual and group effort create an unconscious climate of altruism, where people enjoy the smallest gesture of collaboration (let me help you with this or that — no, don't be suspicious, I just want to help), and waves and waves of small connections are interweaving, emerging as the unique rhythm of this peculiar workspace. Tapping every minute aspect of every single resource, formulating a generous, healthy environment of mutual respect. I'm so pleased to be working with such vital, self-confident, dedicated people. They work for practically no material reward. And this free process is the magic. There is no exploitation — and no real paycheques.

We're just jamming because there's real work to be done in this free laboratory of labour.

I've been charged with participating in any way I can because we are living in the future. And I shouldn't worry if my work isn't exactly the same kind of work my colleagues are doing. Yesterday, I made a couple of unusual, perhaps valuable observations. I was standing on a rocky bank, where the Sable River pours into the saltwater bay of Louis Head, precisely where the fresh water joins the Atlantic Ocean. There were little lion's mane jellyfish swimming in these swirling currents, pulsing exotically with their hanging networks of tentacles indicating the direction of the water currents. I was transfixed for about twenty minutes. Jellyfish after jellyfish would glide by until they slammed into the river's force and would be whipped back toward the sea, like a skydiving parachutist jerked back into the sky as the chute opened and caught the air.

Back down a few hundred metres on the white sand beach was a shallow pool of salt water left behind by the receding sea. The tides have been extreme as the moon has been full. In this isolated tide pool, a lone lion's mane was stranded in seven or eight centimetres of water. It was right side up and pulsing, swimming for its life but going nowhere. From my point of view directly above it, it was pulsing, contracting, and relaxing its gelatinous mass with a rhythm akin to a stimulated heart. It was, for all practical purposes, a heartbeat. As I stood there, my heartbeat and the sound of my breathing and the sound of the calm ocean swelled and locked onto the image of this beautiful, throbbing jellyfish.

And later that night I was sitting in my room, staring at the screen of my laptop with that jellyfish in mind, when I was interrupted suddenly by a different sound coming from my window. A giant silkworm moth was butting its head on my windowpane, flying hard, trying to get to my ceiling light. It was a *polyphemus,* a male with a wingspan of probably fifteen centimetres, a fluffy buff-brown moth with eye spots and all the quivering power of a small bird. It slammed hard into the glass and drove toward the light with all its force, again and again and again, getting nowhere. I stood with my face close to the glass, face to face with the creature, seeing the moth clearly within the backdrop of my shadow surrounded by the reflection of the room set against the rich black of the night. The *polyphemus* just on the other side of the glass, the only real image behind the mirror was alive in the world on the other side of my windowpane. Finally, after a minute, it veered off and the window fell silent and still. I stood looking at my backlit shadow in the window's reflection. The reflection of the ceiling light was white, cool white against the window's flat black; the window was filled luxuriously with the true black of the night.

I wanted to share these experiences with you because these observations are the best work I've done for a while. I contribute them to Recycling the Future in the Hybrid WorkSpace. I realize that it's unlikely that people will have this same kind of experience in Kassel, but it's the best work I have to offer. I wasn't prompted to do this work by anyone or by my need to make a living. No one told me to do this — it was a labour of love. Certainly, there must be room for such work in the free laboratory of labour.

One of the things about the future that interests me is the notion of continuous change and the fatigue and cynicism or perhaps boredom that surrounds a groundless space, where nothing remains the same except that things are always changing and evolving. And with advancing technologies, there are always changes and demands for more and more changes. The frequency of change has risen to the degree that it has become the only constant. I feel this even when I am looking at jellyfish and moths. These magnificent, though humble creatures pay the same price as everything and everyone deemed obsolete or irrelevant or useless by the forces of technological change. Before Darwin, biologists believed that no species of animal or plant could ever become extinct, because that would suggest an imperfection in God's plan. We know better now.

People and the things they do are, of course, also crushed by technological change. The value and nature of labour has changed dramatically. We used to believe the most valuable works were steeped in tradition. Take a walk through the rest of the *Documenta X* and see the moths and jellyfish and birds and delicate wildflowers hanging on the walls. Now we value change and crave change and do our damnedest to shift things in our immediate environments to suit our needs. We were told mid-twentieth-century by the cybernetics researchers that automation and control engineering would eliminate the traditional forms of labour toward the end of the century — and here we are. We were also aware that continuity and tradition would be devalued, change would create value, and the most vital economies would be in turmoil, in uncomfortable social turbulence. Call it what you like, but we've also been hearing for fifty years that as the machines were put in place to take our jobs, we would find ourselves unemployed in traditional terms, and that our greatest challenge would be psychological survival.

With reduced hours of traditional labour, the political and social scientists also pointed out there would be a widening of political participation. And that work and political activity would converge into social engagement and tribalism and all kinds of social currents and hybridism in work, the new work of continuous change. But at least with social and political change we can find comfort in the relative familiarity of the past. Revolutions leap backward to recover what has been lost. The back and forth two-step of political and social change is comforting in its oscillating and pulsing sameness. The relation of labour, forced or free, to political and social change, is a given. The future is well known from this vantage point of memory and hope.

Meanwhile, the moths and jellyfish and birds and wildflowers, the real ones, are disappearing. Extinction is more final than death. But the technologies never seem to be uninvented. There is obsolescence and the machines change and evolve. But there are never any setbacks in the technological realm, their forces never thin out. There is no extinction in the machine world. We, dancing the political and social two-step, forward and back, are caught in the middle, in a world of continuous change.

The frequency of change is increasing on both ends of this formula ultimately and, somehow, this spectrum of natural/political/social/technological phenomena determines the value of our work. We are left to our own devices to cope with the sameness of this condition of change, the numbing redundancy of this shifting ground. My work alone and my work with others are my ways of coping: by generating differences I can relate to and grab onto. So I'm very pleased to be working with you. Thank you for letting me share my work with you and your work. In the future, this cooperative spirit may be all we will have to hold onto: an awareness that our work is valued and that we are working together.

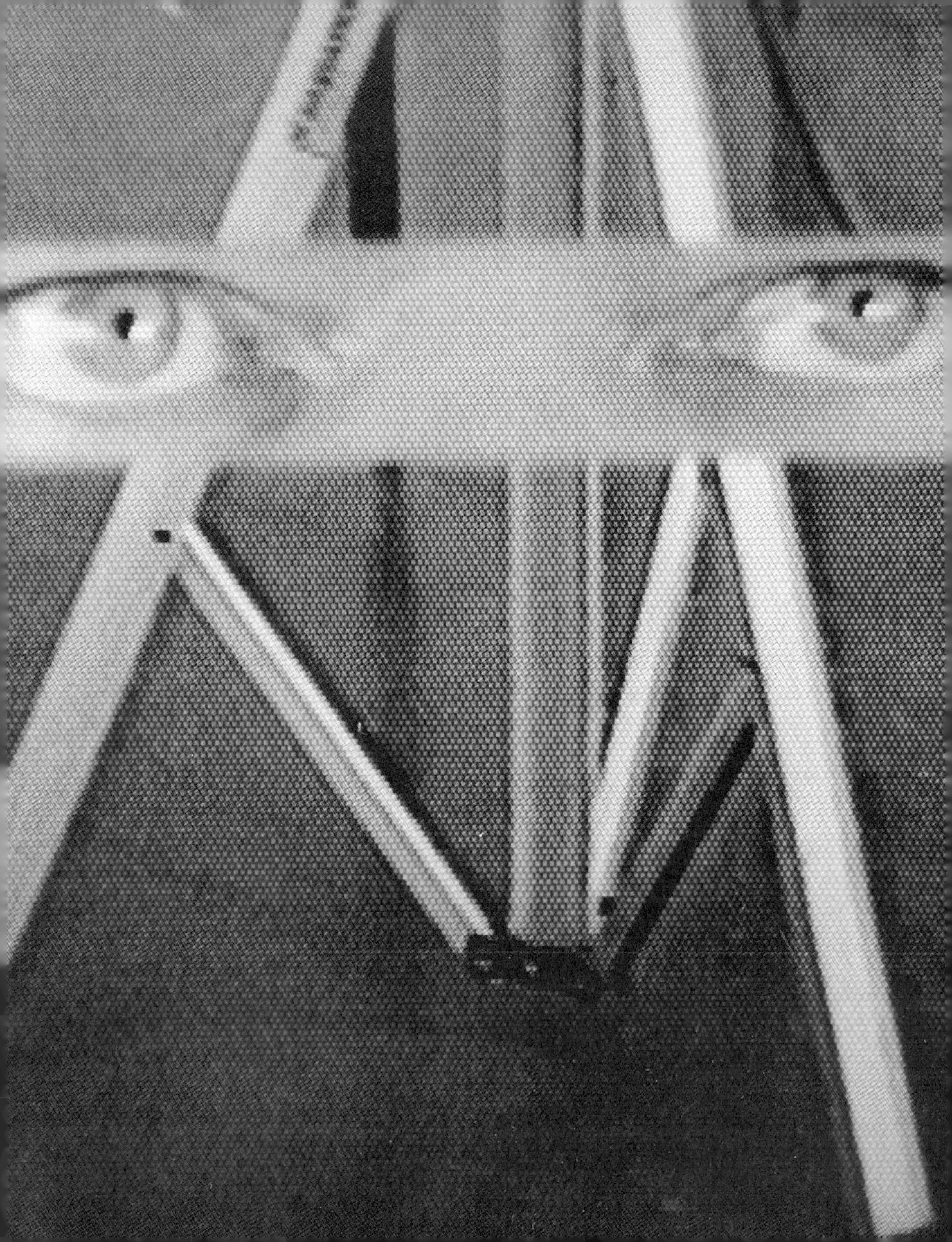

REMNANTS OF PSYCHOLOGY

TOM SHERMAN

Aftermath 1999

The big rage these days is incoherence. Or maybe it is better described as a communications breakdown based on the huge gap between public and private channels? I am not talking about public and private sectors in socio-economic terms. I am referring to public and private communication spheres in terms of relative exposure: the distribution of messages (one to many, one to one, many to many, many to one). There is a huge gap between public and private spheres, and this gap is causing an implosion of coherence and comprehension. The information bomb we are all worried about has already been dropped, in the early 1990s, and we are witnessing its aftermath.

People are alienated from the mainstream, finding they are not represented by the proliferation of industrial television, radio, and print media, while they are simultaneously given license to expand the territory of their private lives through wireless telephony, desktop publishing, home video, digital multimedia, and Internet and Web-based telecommunications. Having spread themselves far too thin in these personal communications media — having revealed way too much private information for their own good — people realize they are overexposed, exhausted, out of material, and facing the abyss between private and public media spheres. Finding their personal information reserves on empty, yet realizing they are still largely unrecognized or undervalued by others, the only clear way to bridge the gap is through acts of violence, figurative or concrete. Symbolic or physical destructive action cuts to the bone in a stifling, discontinuous, inanely superficial universe of symbols, signals, and noise.

There is such confusion at the microphone or keyboard, or in front of the camera. Opportunities for meaningful public communication are squandered by people talking endlessly about themselves. Private communication, paradox that it has always been, is now used as a soapbox for amateur politicians — or just as commonly for the diatribe — the spew of personal venom: look and listen to me, my tongue distributes acidic barbs. No person, no thing can deny that I exist, especially if I aggravate, offend, or hurt them. After multiple transmissions of poison, the diatribist is eventually buried by noise or moderated or shut off by a gatekeeper or a switching mechanism.

But, ah, there are the positive gestures. There are many promoting beauty and intelligence in the void. The wildflowers of the telematic weedbed exist to spread elegance and generosity and love. They act out their instinct to decorate the barren fibre optic tunnels with bright colours and blinking signs and otherwise dazzling design innovations and to provide and manipulate content worthy of experience. Already, there are lists of classic network artworks. Most artists realize the futility of trying to produce ever-green information in a junk culture, and thus the popularity of tweaking (appropriating) or filtering. They transform or filter the economy of abundance (the crap) and provide moments of clarity and order. And they, too, are billed for their time on-line, just like everybody else, for the privilege of inhabiting an artificial, inhospitable telematic space on a planet running out of clean air, water, biodiversity, food, silence, civility. The information bomb was dropped in the early 1990s. It created the space and time we needed to develop our profound incoherence. Just as deregulation has facilitated reformed monopolies, we now have a comprehensive incoherence in the name of cultural diversity.

Nothing Worse 2000

There Is Nothing Worse Than Information That Doesn't Move

One morning I was destroyed by the sound of silence. For all practical purposes, my in-box was empty. I sat motionless in front of my screen. There were copies of messages for other people. There was movement, a busy sense of exchange in the world outside. But I am not a part of this movement. I am frozen. I am static. The world spins around me.

I send out countless messages and only a few come back answered. This morning, there is nothing. I've got information to share, but nobody wants it. I've collected and processed rich fields of data. I've discovered a wealth of information, things other people can use, things that can make a difference. But there are no takers.

I have a site with zero traffic. If I could just get people to visit, I'm sure they would want to hang around. I put all this energy into my information, the way it looks, its composition and depth. There is utility and value in my product. It isn't the same old information you will find in a million other places. My site is different. If I could just get people to see my site is different and to value this difference.

That's my problem. People don't want to be different these days. They want to be part of something bigger, that certain sameness. They want to move like everyone else. They want to belong to the movement, to the spinning world outside. If you want to move, you have to be streamlined. Any difference is a drag. It's a point of friction, of real resistance. Don't dare to be different if you want to move. If you want to go with the flow, you've got to be streamlined; you've got to be smooth.

I don't fit in. The world spins around me. Everything I touch seems to stop in its tracks. I get ideas. I move on these ideas. I make things. They look good. They speak to me. I put these things out there. They just sit. They are stillborn. It's as if they start to crawl, and then they freeze.

I know there are people out there who can use my things. I know I have information, valuable information that people want and need. I just can't reach these people. I can't find them. They don't know I exist. They are somewhere out there beyond the spinning, busy world. Somewhere, out there, there are other people who sit still and watch the world spin around. They are like me. They, too, make information that doesn't move.

Once Living: In a Healthy State of Paranoia 1979

Keep an eye on everything and everyone. Don't trust your mother. Don't trust your batteries. Watch what you eat. Charcoal-filter your drinking water. Learn how to stay out of your body and old age will be a snap.

Remember, even in your silence, you are still more quick than dead. Quiet as a mouse. Common as a housefly. You who choose to exhibit self-control. Keep your lips buttoned tight forever.

Professionally speaking, living in the city, tangled up in media deception. Building careers on hype and long-range projection. Looking into the future for "up to the minute" political forecasts. Everyone should know by now, no one can be everywhere at the same time. No one can be totally aware. Screen your incoming traffic carefully. Be selective, or you'll fry in the overload. These blanket statements are for the victims of the front-page reflection.

More than one of my friends have told me I'm paranoid lately. I always want them to explain what they mean by the word when they stick it to me. Whether or not it is a bad thing to be paranoid should always be up for discussion. I'm not certain it's a negative thing. If I'm paranoid, I'm just trying to avoid the major traps. I know it's a crime to be wound tight with self-restrictions. Don't I owe it to myself to keep my eyes open for an ambush? I simply refuse to ride into the canyon if I know there are bad guys waiting for me on the ledges above. I'm not stupid. But that doesn't rule out my being too smart for my own good. If I felt I had any control over what happens to me, I wouldn't be so pessimistic. I know I can let the whole thing go. I can sit around on my ass, and it will quiet down. I can let it all go. My life will stand perfectly still. I can stop eating. I can stop answering the phone. I can stop paying the rent. I can wait for them to come and get me. I can quit. Right now, I'm absolutely at a standstill. I would rather not continue. Am I thinking about taking my own life? In control, I cease to breathe anything fresh. I shut out the lies. I take possession of my own mind. In the silence of my home. My room is dark and cool. It is summer, warm with the trees sounding so green in the wind, so fresh. My stomach is empty. I don't want to drink. I'm not going to turn on the TV this afternoon. I don't believe any of it. Nothing exists outside my tired body. Slouched in my chair, I've collapsed into the spiral of my dreams. My respiration running down and off. I close my eyes to see what I can see. There is nothing. No colours. No contours. When they come for me, when they finally look me up, I will be moaning blue in rhythms cold without sense. Missing life before my body runs out. Skipping along the path of no return. I don't care one bit. You don't matter to me. Whether you read all of this or not, I don't give a shit. You mean nothing to me. Here I am. Come and get me. Don't try to make me stand for anything. I long to rest, assured of a second thought. I don't see a way out. The clear channel of my silent breathing will no longer support my blind faith in discovery. My desire or invention no longer follows logic. As far as I'm concerned, I have to give up. It's my decision to assume control. I've decided to get out.

It's midnight. It is noon. The mailman is at the door. The phone rings again. It rings and rings. Longer and longer. The sound of the dark green leaves early in the morning. I imagine. The traffic moves. The trains go by. The sun comes and goes. A rain. The moon I imagine crossing the sky. The start of another day. The light floods the room. The cloudy whites covered wet and rolling. The legs are wooden. The hands heavy, the fingers cigarette-stained. I've stopped drinking. The bottles are dry. My urine is yellow. I spit in the toilet. Mouth white. My hair is brown. My beard is grey. My back is asleep. I can feel my chest. My back is there.

I awaken. Here comes the night. I touch my lips. My mouth is open and dry. My tongue is swollen. I see inside. My stomach is yellow. My wrists, my skin is grey. My tongue is white. I'm close now. I'm in control of silence. It's ringing now. I was right to stop running in place. Now I'm running backward, relieved to have hit the slide. Here I go. In the sound of the trees. The shadows playing on the walls. Nothing is lost, nothing gained. I'm sorry to be waiting for you like this. Sure, we could celebrate our consuming interests instead. Excepting, of course, my cowardly exit. What kind of special goodbye would relieve my conscience? I don't know. I don't feel you. I can give you nothing. It has nothing to do with you. Pushing hard, I just can't give. It's too late to pull out. Off this page into darkness. With little pattern to discern. With no lesson to uncover. Dropping off soon. Without a break, I cry. Shit, the tears flow. I didn't want the tears to fall. I tried to hold them back. Dropping onto my chest, the clear tear drops. I hear myself breathing. I feel the clear water of my eyes strike my chest.

I want to. Listen, here comes the end. I want you to know. I feel you see me. I sit; you walk around me. My eyes are dry now. You talk as you circle. I want your words. Talk to me. Walk around me, but talk to me, please. Here it comes. You are in trouble. Listen, open your eyes. Face the music. Cash in on your mistakes. Make money playing around with your life. Be careful in passing. Me, I may register your opinions, indeed. My ears, they'll never close forever. No way, I learned over the years. They can't stop me now. No way, I have learned to reach for the stars. I never thought I would live this long. Must be hell to overshoot the mark. Left standing in space, I'll never believe in another life. You hear me right. My mind is closed forever. I've made up my mind, yet I'm still short of heaven. I hear you thinking, just shy of laughing. In my face, I hear your breath. You told me you thought a life was worth a bundle. You touch me. I'm cold.

More Dead Artists!? 1980

Casualties of the Last Decade, the New Establishment Will Be Dead in 1980

I was talking to the curator of a major Canadian museum and she said to me, "What this country needs is more dead artists." Her words put a lump in my throat. She touched my hand; I had to check her eyes. They were blue-grey. Her whites were clear and bright. She told me it was so nice writing art history when the artists were officially out of the way. It is such a job to fill in the missing lines of their personal-aesthetic profiles. That's the toughest part about creating art history in Canada; unless you include the true natives (the Indians) with their "prehistoric" renderings, it's practically impossible to find enough dead artists to make a believable history. In a believable art history, only the smallest percentage of participating artists break the surface of the turbulent sea of creative obsession. As a result of a wonderfully complex social-aesthetic process of qualitative elevation, a few hopelessly inadequate individuals or institutions are seen to embody everyone's selective sensibilities. Burdened with this cruel responsibility, the highest artists are hated and loved.

Ten Years After

There was once a time when you couldn't trust anyone over thirty. Around this same time, selling out was still an option. If you were making it, you were doing something wrong. Get this: it was then still *impossible* to be anti-establishment in the mass media. There was no time on the air for dissent. Established artists were quick (soon) to be pronounced dead on arrival by the very underground they had emerged from. Meanwhile, the hidden mass of unrecognized artists would continue at a snail's pace their struggle to develop their individual careers in the obscure depths of their underexposed daily activities (beneath the surface of the media). It was a tenuous existence, indeed. Holding one's own in the Western world without much cash, without good plastic; it was so difficult to secure continuous food and shelter. It was very hard to ensure the free flow of tasty booze or the consistent rush of potent-and-pure drugs. It was next to impossible to maintain quality control of the art, let alone these substances of vice, without a second job. The artists who were either too old or too proud to work in the kitchens became administrators instead. Office work was clean and easy, and the hours were good.

The Mega-Alternative Space

In the seventies, a network of artist-run galleries (alternative spaces) was built on the sweat and blood of those punchy young adults who had made the somewhat difficult climb through the sixties and were still alive and kicking. Some members of our mass were direct fatalities of this building process. Many of those who lost their lives were individuals who had simply suffered for too long under the serious delusions necessary to maintain a blind faith in the future. Surely, the State would support the activities of its creative people. Just like anyone else, professional daydreamers need office space and petty cash for supplies. As it turned out, the federal and provincial governments did pick up most of the bills, and consequently, artists were given many years to learn how to manage their own affairs, now as part of the creative fringe of government. Initially, they looked after their own interests, as artists will tend to do. But later, as artist-administrators, they began to take on the responsibility of explaining their personal-social-aesthetic (political) interests to the government before they received the next big cheque. Having once lacked credibility in the competitive world of corporate socialism, hundreds, maybe thousands of artists developed competitive skills through the Mega-Alternative Space (MAS) experience. They learned how to use the telephone and write semiliterate (creative) business letters. And fortunately, these same artists, in their brand new, white-collar government jobs, found the precious time they needed to put into words exactly what it was they wanted to do with their work. They were articulate when they laid out their wishes. Within reason, they were awarded money. And they did their work for years and years. After enough time had passed to warrant an inquisition, this elaborate social-aesthetic (cultural, not necessarily political) construct (MAS) was critically assessed by the uppermost members of our social-aesthetic (cultural, not necessarily political) government. MAS was found to be of sufficient value, at least in purely descriptive terms, to justify restrictive continuation. In other words, MAS would continue with financial support from federal and provincial sources, but with a restrictive ceiling on spending, encouraging innovation within budgets shrinking fast with the inertia of accelerating inflation. And thus, all levels of politicians began to speak publicly on the positive and negative aspects of contemporary art activity and its condition of government "subsidation" (subsidation = subsedation, or a relatively light subsidy, which does not deaden, but is said to stimulate the jazzy unconscious region of the artist's mind through the provision of security as a result of direct funding). So-called high culture had once again become an exploitable political issue. The initial impact and presence of fine art was so magnificently amplified (cleanly and powerfully) according to the classy message of each beautiful fit of art in the mass media. A little art will go a long way these days.

There Was Once a Society of Artists Supported by the Government

What the artist needs is a full measure of respectability in the mass media. This requirement is now tantamount to attaining a respectable position within the artists' community. The artists' community, whether directly supported by the government or not, is a society of artists very tenuously integrated within the greater mass society. For the most part, artists know few people outside their own severely limited society. The artist is an oddball. I'm spouting generalities, of course. But in respect to the discernible patterns of the artist's characteristic behaviour, the mass society has not developed a very considerate mental picture of the artist's mind. The artist is generally seen as a self-absorbed individual; undisciplined, unreasonable, prone to fantasy. People say you never know what an artist is thinking. They never know what their dog is thinking either. And artists make *each other* nervous. Can you imagine that?

Artists Return from Alienation

Finally, the artist, when completely down and out, realizes he or she is just like everyone else in the mass society. Just like everyone else who is down and out, that is. At least the artist *could be* just like everyone else. If the artists in Canada fail to at least elucidate their condition of tenuous integration within the mass society, in the mass media, there is nowhere to go but farther down the ladder. We're deep enough, thank you. Just try to raise money without the mass media. Or attempt to distribute a message farther than your friends without an organ or institution of the mass media. Go ahead. Look and act normal without having the media supporting your idea of normal appearance and behaviour. Try to maintain your sense of self-esteem when you think you are just like everyone else, but nobody wants to know what you think, because they think you are different.

Red Hot Re-Entry into the Mass Society through Death

In reality, one artist is no different than the next artist. And that is about as far as we can go. Having inhabited our tiny world for far too long, we artists will all sooner or later become terminally bored with missing the point of our own esoteric communications. It then becomes necessary for the artist to eliminate the "especially sensitive" aspect of his or her personality. This act might be better understood in terms of a figurative or metaphoric death. A

positive loss of life. Death is one grand process of social equalization. After the artist has dropped his or her superior sense of consciousness, the artist can easily go to sleep at night and dream of dying, just as everyone else does. Everyone dies in dreams. As a matter of fact, in an apparent mass awakening, artists appear to move out into a larger world every morning, searching for relevant work. But first, the especially sensitive artist must die.

The Artist as Fat Cat

Suppose the artist is successful in achieving a better social position within the mass society. Some artists are already sleepwalking; supposing our existing skills could make us useful in the larger world. We might make a living doing our work. We, too, could buy new and used cars and houses and snazzy home-entertainment systems. We could sit around and think as differently as each privileged individual thinks when he or she relaxes at home. Slumping in front of our home-entertainment systems, we would be temporarily relieved of our headaches of cultural apprehension. In a single thought of our vanishing self-consciousness, we might realize that being anti-establishment was simply a state of mind. Good media licks will always make our troubles disappear. I'm having trouble cleaning my head. May I suggest the preventative maintenance of a thoughtful privacy?

Long-Living Healthy Artists

When you live in a beautiful country where everyone does pretty well and everybody is healthy as hell, you will find many people making significant personal contributions to the culture. Some of these contributions will be from individuals who think for themselves. Most believe that the better off the people are, the more personally productive they may be. And if they are in excellent health, they'll live much longer, and their extremely diverse and numerous productions will dramatically increase the volume of our culture in the long run. Micromedia will then be of absolute necessity to cut back the flood of information escaping from the archival tombs of the museums and libraries.

Bloated Egos Bursting with Self-Esteem

It won't be long before everyone will be an artist. If you choose not to become or remain an artist, it will be a very difficult world to put up with. It will be a world where everyone you meet will want to be an artist and will all get a chance to make art. Everyone will say the same things. In such a wonderfully free world, personal identity will be a snap to establish. Very long lives will be much easier to live all the way through to the end.

Artists Kept Alive by Machines

Due to a long-standing very special relationship with high technology, the artists who die the most natural deaths (those who live the longest) will be superbly represented by the lofty performances of the most highly developed machines. This nearly intelligent technology was in fact the key technometaphoric agent responsible for triggering the fiery re-entry of once genuinely egocentric and esoteric artists into the more or less common realm of the mass society. It takes guts to be no more interesting than anyone else. You've got a lot of nerve to sit still, awaiting the eventual flood of publicity as patiently as any machine would, your mind forever closed in the canned work, barring technical difficulty. Listen to the hiss on his lips. Look at the noise tearing through her hair. If you are an artist, your machines are your technometaphoric agents representing you in the sedate subconscious of your most pleasurable future.

The Benefits of Premature Death in a Country Young as This

In Canada, there is presently an almost cancerous growth of art, or at least there is a profound increase in levels of art production. Normally, the harsh economic reality of the land checks the advance of artistic productivity through conventionally fierce mono-occupational competition. But artists today are both superproductive and multi-occupational (multidisciplined). And

they live longer. However abstract their work appears to be, they will eventually be praised by the State for the example of their staggeringly high levels of production. Ours is definitely the culture of the greatest inflation. In the form of completed work, the media mounts as the levels of production continue to rise. Culture becomes nature in the truest sense: as the hills and trees and artists the machines represent. All who produce at such a rate will reproduce in a similar fashion, naturally, in relationships with nearly intelligent machines. The artist-machine seems natural to most of us already. But what about the artist who dies tragically young, before he or she becomes the permanent memory of a machine that can walk and talk? For just as certainly as such an artist-machine would stumble and fall, being so nearly human in respect to grace and intelligence, this same artist-machine's voice must continue to ask one question for us all: Why should a curator-machine's hierarchical order of art history be biased toward a few individuals who meet with tragic death? Remember your dreams. Magic always manifests itself as the resolution of the mysterious. Only those initially deemed as most unfortunate, only those who meet with tragic, premature death will be truly rare in media exposure. Those who die young always cover much more ground in the end. Youth will always recover what the elders have misplaced. Nothing ever changes. Everything is new. The artist has to go to sleep every night, just like everyone else. Close your eyes and listen to your heart pound in your pillow. Listen closely. The ones who die young, their minds are closed forever. And once artists cease to change their minds, the rest is history.

TOM SHERMAN

The Value of Privacy and the Effects of Cultural Engineering 1989

Cultural engineering is the practice of social engineering as it is applied directly to the cultural sector of our society. This text is a result of an increasing awareness of how an economic dependence on funding agencies affects the self-perception and ultimately the public identity of the individual artist. I have developed this awareness both in my capacity as an artist and while serving as a program officer and head of the Media Arts Section of the Canada Council (Canada's national agency for the support of the arts) from 1981 to 1987. I have focused on the media and visual arts because these sectors are my area of professional expertise. At present, I once again work as an artist full-time. It is from this vantage point, the perspective of an artist in Canada, that I offer these thoughts and anxieties.

If we are living in an information society (and many people say we are),[1] and if it is true that there is a new social, political, and economic order based on the exchange of information (rather than material goods), then the individual will have to consume *and produce* information in his or her primary relationship with today's information society. I ask you to consider the individual creative artist as a major, exemplary force in this new society, because artists have traditionally transformed the information they consume into new forms of information in the art-making process and continue to do so today.

If there is an easy way to find meaning in the word *information*, it is as Gregory Bateson once said: "Information is any difference which makes a difference."[2] Important information is information that is particularly significant to us. In this computer age, information is a concept loaded with quasi-religious overtones. It is not unlike science or art in this respect. Despite the flagrant misuse of the word (impenetrable literature on any conceivable subject is often referred to as information), the concept of information is common and means something to everyone. The artist has an important responsibility to contribute meaning to the word information.[3] Gregory Bateson was an important innovator in this regard. He certainly was an artist to the degree that he refused to tolerate professional restrictions as a thinker.[4]

The individual artist consumes and produces information on a daily basis in his or her work. The process of making art includes discovering significant order (or disorder) in reality, which subsequently immerses the individual in the problem of stating clearly the precise nature of his or her personal relationship with the world. Art is first and foremost information produced by individuals. This observation is not to discredit collaboration or to discourage social interaction, but only to claim that there are distinguishing characteristics within an individual's psychological domain that add up and have inherent value. In some instances, personal information refuses to bear the burden of the thoughts and actions of others. Authentic personal information is not authored by committees or boards of directors. The social relationships that influence information produced by organizations give institutional information very different characteristics. In terms of influence, institutional information is increasingly the basis of authority in the late twentieth century.

Organizations engaged in the production of authoritative decisions about people base such decisions on personal information.[5] Much has been said and written about the invasion of privacy, the erosion of personal autonomy, over the past three decades, especially in the United States. Perhaps this attention is due in part to the horrific abuses conducted by Senator Joseph McCarthy's officials in the fifties, or the dirty tricks attributed to Nixon's administration around the time of the Watergate break-in. Abusive invasions of privacy can also occur in Canada or any other nation where organizations consume personal information as the raw material from which they formulate authoritative policy. Despite society's concerns over diminishing privacy and eroding individual autonomy, artists exhibit a strong tendency to produce and disclose personal information voluntarily.

Before attempting to characterize the value of privacy, it is important to examine the conceptual relationship between privacy and personal autonomy.

Privacy is that aspect of identity that is personal, not publicly known. It contains information about the self that is confidential to, that belongs to, and is the sole concern of, the individual. It is closely related to the individual's personality and is the perceived or preferred concept of self. Personal autonomy depends on the individual's ability to control how and under what conditions the private self is revealed to a public world.[6]

A fundamental distinction may be made between *aesthetic* and *strategic* privacy.[7] Aesthetic privacy is maintained by restricting personal information as an end in itself. In such cases, disclosure would be inherently embarrassing or distressing. (For example, to hide from others acts of excretion, sexual intercourse, or profound emotion would maintain this form of privacy.) Strategic privacy is attained by restricting personal information as a means to some other end. There are instances where privacy facilitates the pursuit of other interests. For example, a father might wish to conceal ambiguous news about his health from his family until matters can be clarified. Or an employee might wish to conceal from a current employer, plans to seek a new position elsewhere.[8] In the arts context, an artist might choose to sit on an idea or image until he or she has the means to mount an effective public exhibition of a work of art. Although both forms of privacy are essential to the well-being of the individual, the necessity for maintaining aesthetic privacy is directly related to the instinctive shyness common to other forms of animal life. Without sufficient aesthetic privacy the individual will simply fail to thrive. Strategic privacy, on the other hand, though not essential to mental and physical health, is constantly challenged by centralized, authoritative organizations. In the case of many individuals, particularly creative artists, the line between strategic and aesthetic privacy is thin.

Few individuals have the option of avoiding relationships with record-keeping bureaucracies. To do so is to forego not only credit but also insurance, employment, medical care, education, and all forms of government services to individuals.[9] The artist, of course, must submit personal information to federal and provincial agencies in the process of competing for grants. Cultural agencies, like all authoritative organizations, are consumers of personal information. Applications submitted in competition for public money require the disclosure of personal information. At the outset, an organization's request for information may seem to be a harmless formality; the application forms usually request the following: name, sex, marital status, date of birth, citizenship/resident status, address, telephone number(s), current employment, an updated curriculum vitae, a detailed statement of intent, a detailed projected budget/including schedule of work, names and tasks of potential employees, a complete travel itinerary for the duration of the grant, notice of other grant applications/past and present, and personal references.

Upon receipt of this information, a file is opened on the applicant, and from this point, officials in the organization become professionally responsible for systematically monitoring the relevant behaviour of the person represented by such a file. This bureaucratic responsibility is called surveillance. This term is used here in a neutral sense to indicate monitoring for all sorts of purposes, potentially benevolent or coercive. The purpose of all surveillance, in the broadest analysis, is social control. Social control is any means of influence by which a person or an organization seeks to render other people's behaviour or circumstances more predictable and more acceptable.[10] All things considered, the enforcement of norms on a pool of artists applying for financing to complete personal projects would probably never be as uniformly restrictive and unforgiving as a screening of individuals applying for the possession and privileges of a major credit card. This is not to say that applicant-artists would have the option of maintaining anything other than a communicative, cooperative relationship with the bureaucracy and all those entrusted with its authoritative decision-making responsibilities.

Personal information files tend to take on lives of their own. The written word, or the instantly retrievable computer text file, represents the ideal medium for bureaucratic action. People who work in bureaucracies must be able to show, to their superiors or to inquiring outsiders, that they did the right things under all circumstances. This requirement places a premium on documentation. Hence the well-known reflex of creating a written record of every action for which the official could conceivably be held responsible.[11] As each file grows, the organization that consumes personal information also *produces and provides* information, first for its own use and potentially for unforeseeable uses. Although the unauthorized transfer or disclosure of personal information is prohibited by law under the *Privacy Act* in Canada, it is still common for personal information to be shared by "families" of authoritative organizations.

Such a family might consist of communicating arts councils, government departments, public archives, universities, libraries, and museums. The Canada Council, for example, turns all its files over to Public Archives Canada eight years after the year of each file's creation. These files are then available to the public for research purposes.[12] This fact should offer some consolation to those concerned about the apparent dearth of documentation of contemporary art history. At the same time, it may come as a shock to many artists and arts administrators who assumed they were operating within a system governed by strict, time-honoured confidentiality.

It might also intrigue some to know that the Canadian Security Intelligence Service (CSIS), has virtually unrestricted access to personal information files if its officials feel there is reason to believe an artist or his or her associates pose a threat to national or international security. In other words, if CSIS wants personal information on an artist, it has the legal right to examine files held by public agencies. So a family of authoritative organizations might include two or three arts councils, a couple of government departments, and an indeterminate number of cooperating secret service organizations. There can be no doubt that excessive concentrations of power are forming in this society. The balance of power is shifting increasingly in favour of bureaucracies, both public and private, at the expense of individuals and those of less organized sectors of society.[13]

Authoritative organizations transform personal information into social, political, and economic order. In terms of arts funding, cultural agencies are somewhat analogous to credit bureaus. Grants are advanced in much the same manner as loans against the submission and collection of personal information. It is difficult to get a grant without an established a grant history.

And there are always limits to the generosity of any system. The granting authority, through the decisions of a board of directors, a staff of official employees, and a host of external assessors, determines how far to extend the rewards of the system and will attempt to tailor the amount and nature of the privileges to its reading of what is safe for the individual artist concerned. What is "safe" is what is deemed an acceptable level of public support per individual artist by those who govern the system.[14] When this safe or acceptable level has been exceeded, it standard practice to withdraw "abusers" from the system. If a nation's artists were to become highly dependent on public funds distributed by a single authoritative organization, it is conceivable that such acts of corrective maintenance would literally take certain artists out of circulation.[15]

If there is one glaring omission in this muse so far, it is the absence of speculation on how the artist's primary act of information production, the art-making process, is affected when the artist is involved on an ongoing basis with an authoritative organization such as an arts council. Although most individuals in a society choose to remain anonymous to those outside their family, friendship, and employment circle, the artist is commonly engaged in the extension of his or her influence through exhibition. There is also a clear pattern of psychological transgression in the professional behaviour of the individual artist. There are periods when the substance of an artist's work typically consists of formal adaptations of blatantly personal material.

Definitions of self can, of course, be rendered in any medium. But one need only look back to performance and video art in the seventies to recall works of art that consisted primarily of the disclosure of what would normally be considered private, personal information. It is not entirely coincidental that performance and video art were supported generously relatively early by arts councils, and that these art forms to this day are supported, virtually without exception, by authoritative organizations with public funding. There are also indications that artists working in performance and video often find the rate and extent of their public disclosures accelerated and increased by the system that supports their professional activity. In other words, we have a dangerous situation where creative activity that manifests itself in the public disclosure of personal information, often through personal appearance and autobiographical narrative, is supported by authoritative organizations that make decisions based on personal information. Applications for performance and video grants are based on personal performance on video. This reality puts individual artists in a sometimes positive, more often negative, feedback system characterized by high risk; funds for subsistence, production and exhibition must be secured through the strategic release of personal information, with the stated intent to make similar disclosures in public exhibition.

Although there are definite, problematic similarities between the process of applying for grants and disclosing personal information to an audience, the implications of such an analogous relationship were overshadowed in the seventies by the vitality and power of personal work and its disturbing effect on audience. The arguments against this kind of work were usually emotionally charged by the indignation of the audience. The artists worked hard to engage anyone who would pay attention. At least part of the strategy was to embarrass the audience with the awkwardness of the disclosure. Those who refused to accept the "life is art" philosophy were adamantly bored to tears with each successive admission. In those days, the formal device was clearly the personal narrative. And although there was an immense body of work in literature and theatre that exemplified a refined sense of personal narrative, all such interdisciplinary art history was ignored by most, except for a few serious individuals, in favour of the irresistible cause: the reformation of popular culture. Most artists never looked back.

For those still interested in developing personal narrative as a form, the video medium and the computer, which complements and controls it, offer the individual artist the means to document a personal, private reality with the preferred tools of authoritative organizations. Video is the ultimate technology for behavioural surveillance, and the computer's vast memory and information-processing capabilities, linked with telecommunications, provides formal and technical compatibility with all record-keeping bureaucracies. These ubiquitous technologies of memory facilitate the disclosure of personal beliefs and the detailed documentation of lifestyle and routine. The artist records his or her own personal admissions in preparation for a formal public disclosure. The following is a constructed, composite statement, representing many points of view, which may serve to demonstrate the personal nature of such public disclosures by artists:

> Sometimes I wish I was gay, I'm a feminist, although I'm not a socialist or an environmentalist; quite frankly, I wish the whole world would be destroyed by a thermonuclear war; I'm a voyeur and an exhibitionist, a foot-fetishist, an adolescent at heart and an anarchist deep down inside; I'm a heroin addict and I commit crimes against others to support my habit; I'm heterosexual but there is nothing I like better than dressing up in my mother's uniforms and wearing heavy makeup; I'm all alone and desperate for companionship, afraid of dying naturally, sexually misfunctional, not satisfied with my life; I hate my body, feel inadequate intellectually; I wish I was living in the eighteenth century.

The artist polishes and records these disclosures in a variety of forms and media, and then actively seeks opportunities to make these private admissions public.

Artists are not the only individuals in our society to exercise their rights to disclose personal information (just check out the confessions of the masses on any late-night radio talk show), but many artists have a need to disclose what would normally be considered private and personal. Although many intelligent individuals operate under the assumption they are being monitored surreptitiously, the artist takes the possibility of being the focus of unapproved observation and turns it into a performance situation. The victim of surveillance emerges as an exhibitionist. But as confident and self-assured as the artist may be, everyone has an inner hidden self, a protected zone of privacy, inside which he or she conceals ultimate secrets. In reality, an artist's disclosures are often diversionary tactics. The individual may be "arranging" the nature of his or her disclosures, or be dispensing fictional "secrets" in an attempt to create a desired impression on others. Such strategic admissions are part and parcel of a technique called impression management.[16] The self can be, and often is, constructed. Most artists prefer to create and maintain their own persona.

From the perspective of the official of an authoritative organization, the challenge is to find order in the disorder of the multiplicities of human biographies, characteristics, abilities, and intentions.[17] The bureaucrat desires to predict, monitor, and control these sources of disorder. Documentary information on otherwise obscure individuals provides a source of strength to authoritative organizations. If a particular piece of personal information enables a decision maker to separate individuals into categories that correspond, on the average, to the sought-after characteristic, the piece of personal information will have economic value, because possession of it increases the chance that a given decision will produce the intended result.[18]

A complex story is made up of small details. The personal history of an individual can be construed by piecing together a string of seemingly insignificant transactional events. How does the individual handle money? Does the individual travel? How often and where to? Where does the individual reside permanently? Does the individual change addresses frequently? With whom does the individual associate? Does the individual honour formal agreements? If an organization assembles detailed notations about everyday events in the lives of a significant portion of a particular professional sector, then the organization is undoubtedly building a valuable base of knowledge about its information providers. The existence of a significant body of computerized transactional information allows an organization to develop profiles of expected behaviour. Transactions generally thought of as isolated and unrelated events can leave a trail of behaviour that can be stored cheaply, in perpetuity.[19] Memories fade, but information retained by technology has permanence. A clear depiction of a set of bad habits may lead to outright discrimination. It may be impossible for the individual to make a fresh start.[20] In recording an individual's past misdeeds and shortcomings, modern bureaucracies may erode a desirable element of forgetfulness in social relations.

The artist who contributes personal information on a frequent basis to an authoritative organization, whether in competition for public funding or in an advisory role in assistance to an organization's decision-making process, should be cognizant of the contribution he or she is making. These personal files, initially self-authored by the individual in a series of autobiographical sketches, accrue over time into what could be thought of as individual privacy reserves. In-depth personal histories are assembled from a series of discrete (and discreet) information exchanges. Personal information is exchanged for an organization's literature (information sheets and application forms) and whatever correspondence and messages its officials deem necessary to document an official information transaction. On occasion, the organization's official message is money, a grant of public funding, and the subsequent public disclosure of this decision and expenditure. Perhaps these personal-institutional information exchanges should all be characterized as economic transactions in a developing information economy, whether or not a cheque is ever issued? After all, every single information transaction, which consists in part in the submission of personal information, is of value to the organization (and ultimately to the society) as it contributes fundamentally to the establishment of performance standards for, and contributes to an expected, normative behaviour of, the individual artist.[21]

The basic principles of the developing information economy include the universal maxim: order has value.[22] Benchmarks are necessary whether one is putting a price on real estate (a painting, sculpture, or installation) or establishing the relative value of the beliefs, behaviour, and performance of the individual artist. Without a highly developed sense of an expected or probable personal philosophy, behaviour, and performance of the individual artist, it would be impossible to recognize new, socially valuable ways of thinking and acting. As diverse, disorderly volumes of personal information are set in order by the information-processing methodologies of professional bureaucratic officials, the main objective of the authoritative organization becomes perfectly clear: the mission is to encourage work that is good for society. In a nation like Canada, where authoritative organizations play such a major role, it could be said that the production and distribution of art is governed by an economy of social value.

Those insistent on social reform will find little satisfaction in the current state of affairs, but these people can sleep at night, knowing there simply is not the capital investment potential within Canada to tip the scales toward a virtual market-driven aesthetic. In other words, the art scene in this country will likely stay much the same as it is at present. There will be a mix of some overtly political art and lots of vaguely apolitical art. Almost all works of art in this country will continue to be created "for the good of society" by artists involved in ongoing relationships with authoritative organizations. Very few artworks are made without ethical or moral deliberations. Artwork in Canada that appears to be contrived to represent a pure state of ethical and moral vacuity is more likely to be symptomatic of a widespread psychology of obscurantism among artists in Canada. Some artists deliberately code their artworks for the international commodity market, but most of what appears to be functional commercial coding for the attraction of investment capital serves far better as an effective cover-up for the disclosure of personal information.[23]

Strategic tactics designed to obscure what is considered to be private information are most evident in object-based art (painting, sculpture, and installation), which does not involve the artist in personal appearance or autobiographical narrative. Visual artists who make object-based art, distanced from the self by physical materials and often obscure aesthetic codes, appear to be affected by the bureaucratic realities of cultural engineering in a different way than those prone to autobiographical disclosure. Making written submissions of personal intent and rationale to granting authorities appears to be offensive and painful to these individuals. They rarely decide to disclose details of their private lives in the act of public exhibition.

Before authoritative organizations like arts councils evolved into such voracious consumers of personal information, the critic and curator were once the primary speculators on the value of privacy. Their professional affiliation with publications and museums, which are different kinds of authoritative institutions tied more fundamentally to audiences, dictates a different kind of relationship and behaviour with artists. The standard method of information gathering is the taped interview, conducted in the privacy of the artist's home or studio. The interviewer almost always assures the artist that the tapes are being made for research purposes only and that the recorded conversation, in its unforgivingly precise electromagnetic form, will remain private. Critics and curators are, of course, generally viewed with less suspicion, as their expression of interest in an artist's work, beliefs, and lifestyle is usually perceived as a blessing, a chance to explore the public potential of the artist's work.

It is fortunate indeed that most bureaucrats in cultural agencies and many critics and curators are sensitive and ethical people. Most professionals in the arts who make their living handling personal information are aware of the importance of respecting the privacy of the individual artist. The official of an authoritative organization must stay informed by scanning and collecting information, not only personal information submitted by individuals themselves, but the oral histories of the individuals contributed by others in the field. Careful listening and good judgment should govern this research. Gossip and misinformation should never, under any circumstances, be documented as fact in a file. Eavesdropping must remain forever unacceptable as a surveillance technique in such noncriminal investigations. The most objectionable techniques for encroaching on individual privacy must never be tolerated in the arts sector: i.e., electronic surveillance, polygraph testing, personality testing, subliminal messaging, and the manipulation of personal files.[24] In the case of the latter, it is a responsibility of the artist to reveal to others all hidden messages emanating from authoritative organizations and to resist the temptation to manipulate audiences by subliminal invasion of the protected zone of privacy.[25]

It should be clear that there are inherent dangers in the practice of cultural engineering.[26] Authoritative organizations operating in the cultural sector do not play a neutral role in determining the future status of the artist in society. Many questions have to be asked. Do systems that collect personal information inevitably bring about changes in relations between centres of social power and the peripheries?[27] If the normative behaviour and performance standards for the individual artist are determined in and for the centre, what does the authoritative organization offer the individual at the periphery? Although the above two questions focus on regional disparities, the periphery can also be broadened to include all those marginalized by the system: minorities, including women and cultures (Native Indian, Asian, African, Caribbean, Latin American) other than the dominant English and French Canadian cultures. Furthermore, will a large, comprehensive system, which accumulates personal information from so many, effectively deny innovation and dissent by defining an acceptable level of unorthodox or radical behaviour? And, finally, from the other side of the equation: if order has value because it enables new forms to be created out of old forms,[28] can brand new kinds of artists be created out of run-of-the-mill artists through the personal information-processing dynamics of effective, strategically directed authoritative organizations?

The morass of psychological and social horrors touched on in this writing may distract its readers from the author's main interest: that is, there is essential value in the private reality of the individual, and, however threatened, this value of privacy seems to be a key aspect of our developing information-based society. Artists, individualists among individuals, have a real interest in giving definition, and therefore value, to private matters. Unfortunately, by offering details of their personal world, often elucidating the unspeakable, artists fall into a conundrum of paradox, forwarding gestures of self-destructive generosity to the information-processing machines — those engines of social reformation, the authoritative organizations themselves.

The audience is thus served a course of well-screened disclosures and cover-ups, both within and beyond the bureaucracies' range. After all, the system is designed to create stability by doling out proportionately more negative than positive feedback. Some artists go public with painfully revealing disclosures; others put forward empty vessels devoid of all personal background. Although authoritative bureaucracies, with their technologies of memory (video and computers), do not strictly control the nature of these mixed emissions, they do in the long run increase the sameness, the predictability of both naked and cloaked disclosures.

But have faith. Against the odds the individual artist is still capable of investing his or her private thoughts in new plans of action that will characterize and enhance the value of privacy for everyone.

1 Two good studies of the information society are Daniel Bell, *The Coming of Post-Industrial Society: A Venture in Social Forecasting* (New York: Basic Books Inc., 1973) and Simon Nora and Alain Minc, *The Computerization of Society: A Report to the President of France* (Cambridge, Mass.: MIT Press, 1980), originally published as *L'informatisation de la société* (Paris: La Documentation Française, 1978).

2 Gregory Bateson, *Mind and Nature* (New York: Bantam, 1979).

3 Tom Sherman, "The Rabbit Theory of Data Transformation," in Manon Blanchette (ed), *Un aspect différent de la télévision* (Montreal: Museé d'art contemporain, 1982). In this essay, I attempted to integrate the information theory of Claude Shannon ("A Mathematical Theory of Communication"), published in 1948, with my philosophy of video aesthetics. Shannon's information theory, to quote from Hiroshi Inose and John R. Pierce, *Information Technology and Civilization* (28, New York: W. H. Freeman and Co., 1984), "says nothing about the meaning or content of information. It deals only with the amount of information, measured in terms of choice in the generation of message and uncertainty as to what message will be received."

4 Gregory Bateson, *Steps to an Ecology of Mind* (New York: Balantine, 1975).

5 James Rule, *The Politics of Privacy* (New York: New American Library, 1980), 49.

6 Arthur Cordell, *The Uneasy Eighties: The Transition to an Information Society* (Ottawa: Background Study 53, Science Council of Canada, 1985), 94.

7 Rule, 24.

8 Ibid., 49.

9 Privacy Protection Study Commission, *Personal Privacy in an Information Society* (Washington: U.S. Government Printing Office, 1977), 16.

10 Rule, 47.

11 Ibid., 49.

12 The Canada Council stores all of its files on individuals and organizations at its premises (Ottawa) for three years from the year of each file's creation. After three years, all Canada Council files are moved to and stored at Public Archives Canada (Ottawa), where they remain property of the Canada Council for five more years. All Canada Council files become public property eight years after their year of creation, available for research to the public at large (subject to protection by the *Privacy Act*) through the services of the Historical Branch of Public Archives Canada.

13 Klause Lenk, "Information Technology and Society," in Guenter Friedrichs and Adam Schaff (ed.), *Microelectronics and Society: A Report to the Club of Rome* (New York: New American Library, 1983), 271.

14 Rule, 48. The ideas in this paragraph are based on a general description of how credit bureaus use personal information to ferret out high-risk individuals requesting bank loans or other forms of credit. Not coincidently, this manner of thinking is often prevalent in boards of directors earnestly trying to improve the effectiveness and fairness of cultural agencies that provide arts grants.

15 Ibid., 48.

16 Cordell, 94, in reference to W. Lambert Gardiner, *Personal Data Banks and Personal Autonomy* (Ottawa: Science Council of Canada, 1980).

17 Rule, 188.

18 Roger Noll, "Regulation and Computer Services," in Michael Dertouzos and Joel Moses (ed.), *The Computer Age: A Twenty-Year View* (Cambridge, Mass.: The MIT Press, 1979), 261–62, in reference to A. Michael Spence, *Market*

Signalling (Cambridge, Mass.: Harvard University Press, 1974).

19 Cordell, 75.

20 Lenk, 276.

21 David Burnham, "The Rise of the Computer State," in Lewis Lapham (ed.), *High Technology and Human Freedom* (Washington, D.C.: Smithsonian Institution Press, 1985), 143. Burnham's writing on expected behaviour has been used to formulate this paragraph, and all successive speculation on the anticipation of artists' beliefs, behaviour, and performance.

22 Jeremy Campbell, *Grammatical Man: Information, Entropy, Language and Life* (New York, Simon and Schuster, 1982), 41–2.

23 The ideas in this paragraph developed as a result of written correspondence with Roger Mayer of Providence, Rhode Island, as perceptions of American and Canadian art economies were exchanged.

24 Rule, 73, in reference to Alan Westin, *Privacy and Freedom* (New York: Atheneum, 1967). These generally objectionable techniques are listed in five case studies of policy making in matters relating to privacy.

25 A. E. Dick Howard, "Privacy and Autonomy in American Law," in *High Technology and Human Freedom*, 35. The concept of a "protected zone of privacy" has been used in the legal defence of an individual (*Stanley v. Georgia*, 1969).

26 Tom Sherman, *Cultural Engineering*, Willard Holmes (ed.) (Ottawa: National Gallery of Canada, 1983).

27 Rule, 119.

28 Campbell, 41–2.

TOM SHERMAN

A Statement from Inside the Cultural-Industrial Compound 1981

I want to become more productive. I can't effect enough change by myself. I must make a difference. My work is too important to be ignored. I must find a way to extend myself, to influence others. I know they are out there. I can feel them. There are others like me. I can't sit on what I have any longer. I must come out with it. The way things are going, I can't stand the thought of remaining inside. I have to make a difference. I can't stand the thought of being overlooked. I'm begging for an audience. I have to reach the right people. I can't find them without you.

I need your help to find them. Through you I will find them. I know I'm pressing. First, I must reach you. I have to get to know you before I can advance. Before I can touch them, we must recognize each other. I see you on the street. I begin with you in a crowd. You stand out. I know you are the one before you notice me. I am looking for you. You don't know I'm coming.

I'm presenting myself. I'll be clear — I'll stand out. I'm waiting. I'm being patient. If there is one thing I've learned over the years, I know I must wait. I want you to see me in your own way. I want you to find me yourself. I'm desperate to reach you, but there's time. I know there is time, but I feel I must connect with you soon. You are the key. You will be my introduction to them. I know you are the key. I can see it in you. You're slowly moving into place. You are coming around. I remain motionless. We are much closer. We are closer to the same. I am watching you change. It's my advantage. I respect you. Take your time. Be precise in your observations. Look for me where nothing else matters. I'll tell you a lie to get your attention. I'll tell you I know all about the others. I know what they are looking for. They are looking for something inside. I have something hidden away for them. I offer you everything.

Lift me out of this text. Extract me with your concentration. First, I will speak to you about the others. We must share the others through our differences. The others will come between us initially. We will see them clearly through each other's eyes. Turn around, I can't wait any longer. I'm here for you now. I've set a trap for you. I'm not interested in the others at all. I've set this up for you. Turn around, I'm behind you. I represent you. In a way, you depend on me. In a way you don't understand yet. I'm waiting for you to turn around. I can't wait. Our time together begins here.

We are together in the same thought. You are beginning to sense my direction. I am focused on your image. All of who I am is invested in your image. I have fallen for you. I hope you will recognize my investment. I am coming up behind you. I support you. I know you. Turn around. I can't wait any longer. I want to touch you. But that would be too easy. That would be a cheap mistake. It wouldn't last. You have to find me yourself. You must eliminate the others. As you clear away the others, I am moving around you. I am cautiously manoeuvring into better position. At this moment, I am moving under you. I am under you. I want you. I am under you because you are all that matters. I am waiting for you to make a move. Do something. Seize control of the situation. The circumstances favour your immediate approach. Take me. I face your rejection. You sense I am under you. You recognize me. If you can see me, why can't the others? Why can't I reach them? Can you explain how you see me? Tell me with your eyes. Take your time. Your face has changed in the last few moments. You are different now. What is it I see in your eyes? What is it? Easy. Keep me under. Your face has changed. Your face is their face. I see their eyes. Through your eyes they must see me. They see me beneath you. Through your eyes, they are aware of my presence.

Now that I have their attention, you are expendable. I'm sorry, but I've been receiving their signals through you. Your transparency is your only saving grace. They are amazing. Their eyes deliver so many stories. I love their mass. I'm drawn to their texture — their totally involving surface. You remain a ghost. They are your new flesh and bones. They are shaped by your form. For me, they have become you. They are you. I know I've just finished dropping you cold, but I've realized a few things, and I want you to know. While

I was apparently intensely involved with their mass, your singular form was most attractive to me. I could never see them for who they are. I could only see them in relation to you.

Before you write me off — I can change. Have faith in me. I can change. I want to recognize you for who you are. I've decided they are far less interesting to me than your phantom. But your fleeting image itself is not enough. I've lost touch with both you and them. They would simply not be deep enough in the long run, and I'm afraid I've lost my grasp on who you are. If I were on television, if you could imagine that, I would have my own channel with no regular audience. I would be alone on the screen. Most of the time I'd be talking on and on to myself.

I have a channel's time to fill with no reason. I sit "on camera" in isolation. This display is pitifully empty. There's nothing here, and there's nothing out there. I'm opening myself up all the way, but there is no flow. I'm on television — I've got my own channel. I could be reaching millions. I feel only you. I sense you are out there. I first noticed you were on my channel in the mornings. You would come to me while I was washing my face with hot water. And I would feel you later after lunch when I would put my feet up on the coffee table. I found I would break the connection if I went to the window to look for you. It was difficult to act impulsively without breaking the connection. When our channel broke down, it was hard on me. I would feel painfully empty. I would freeze inside. Then you would return — I'd feel that subtle buzz behind my eyes. You are on the other side of that camera. You are the someone who is watching. I raise my eyes to the lights, exposing my throat. I sense your reaction. I turn my back on you and lower my eyes. I guide my hands to my hips. I study the pressure in the soundproof room. There is no one here. There is nothing but a silent static field of hot studio light. That loaded expanse of bright white light.

I clasp my hands behind my neck and spread my chest to the camera. There is nothing coming back. I stretch open farther. I don't have a plan. I've misread the pattern. You are gone. I'm stretching myself too thin. I'm not strong enough to sit tight in my own weakness. I'm trembling with embarrassing fragility. I must bear down to find myself. I'm pulling in my feelers, all my queries. Once more, I'm my own total audience; 100 per cent inside again. I ask you to please accept my chronic introversion.

Again, I face the possibility of your rejection. I can only hope that my collapse will interest you. If nothing else, you may watch me shrivel up. I will withdraw completely. My expressions will all be grotesquely distorted by this unavoidable, irreversible compression. I will tighten and blacken under my eyes. I will soon be unrecognizable, dark, and unknown. I will completely shut down my physical body — I will black out, and I will inevitably face the critical problem of heat loss. I will dry out. I will cool off. As I lose my body, I will progressively become transparent. Exposed but shameless. My being contracts to find an acceptable level of presence within the context of this station. I concede to the colour of the room. I become beige. I am perfectly motionless. I am as wooden as the table I sit behind. I am as static as a word. I have become the word *nothing*. My life is worth nothing. I disappear. This is self-destruction. I ask you who will be next? Enter the studio of Channel 107. The self-destruction chamber. I am inside. I am nothing, and there is another full day's programming to fill. I'm scheduled to be on "live" from 6 a.m. to midnight. I program the whole channel with nothing but my presence. Good morning, this is Channel 107. I am nothing.

What is the value of the individual artist in a society where culture is discussed in industrial terms? In the spirit of evolutionary self-destruction, I offer myself to the dream state. There are ants in my nose and my tongue is cold. The machine watches and records me exactly as I appear to be.

If there is a lie, the camera tells it. If we cannot believe the recording mechanism, who can we trust? I ask you to believe in me. I am behind you. I represent you. I am under you. I've got ants in my nose and my tongue is cold.

Dead Faces 2001

For the past four months, I've been communicating with the main office via Webcam and voicelink. It has been interesting to supplement our normal e-mail links with pictures and more expansive voice conversations. We still keep things tied together with text. Written memoranda still rule in terms of documenting exchanges and getting things done. But Webcam talks are very useful for reports and planning. Conference calls make more sense with screen support. I've enjoyed getting a broader sense of my coworkers, the way they look and talk.

I've also noticed a lot of corny, forced humour that seems to be generated by these Webcam encounters. In a situation where four or five of us get together on a shared screen to develop a strategy, it's crazy the way everything is put forth with grins and smiles, no matter how serious the meeting's undertone. There seems to be an unwritten code of behaviour that forces everyone to be silly and predisposed to belly laughs at the expense of everyone and everything. This goofy, obnoxious code of conduct is particularly glaring to those sitting in on these meetings off-camera and mute. Our corporation has a transparency policy. The times, addresses, and agendas of all Webcam meetings are published in advance on the corporation employees' home page. Everyone in the corporation is welcome to look and listen in on every meeting, should the agenda be of interest, or if they want to get a sense of what it would be like to work with particular individuals at a later date.

Personally, I find the time just before and after these Webcam meetings to be the most revealing moments in the life of the corporation. If you look at the daily meeting schedule and log on ten or fifteen minutes before a meeting commences, you will find that some of the participants sign on early. They are usually sitting there in their cubicles dealing with on-screen data or attending to their personal e-mail or telephone conversations. It always strikes me how dead their faces look before the meetings start up and they put on their perky corporate personalities. It's the same after a meeting when they're working alone again.

If people would look carefully into their video mirrors, they would notice a certain deadness in their faces. People sitting at computers for long hours have this extra gravity in their expressions, especially in the muscles around their mouths. When I look in on people getting ready for a meeting, or just after, it is striking how they appear emotionally mute. This vacant, drained appearance triggers my own depressing introspection. I've become aware of my depression through the numbing fatigue around my mouth, the way my face looks and feels heavy when I have that sinking feeling.

This is what I've learned from working in an organization linked by Webcams: I've learned that we're networking a certain level of depression. I don't believe the machines are draining us, but there is something desperate about the explosions of laughter that erupt when we're interacting on camera. When I feel numb and emotionally exhausted, and look particularly drained around my mouth, I usually try to go outside for a walk, to get some fresh air. Or, at least, I stand in front of a window. I find it helps to go outside or someplace where I can be outside in my mind.

TOM SHERMAN

Suggested Sway 1998

From my couch, I'm worried
I have a fever
But I'm still my own person
It started with my parents
They didn't know what they wanted
They just wanted me to cooperate
They sent me to school
and church
I had friends
but the closer we got — the more demanding they became
We grew apart over the years
When I was old enough, I quit school, and went to work
In my second job, I got fired
I admitted I was responsible for human errors
The police came
I was on television and in the newspapers
Finally, I shut it off — I terminated the flow
My doctor sent me to a psychologist
He told me I have a low information drive
I think it comes from long stretches of living alone
With no regular hours
From not leaving my apartment for days — maybe weeks
No television, no radio
The mail comes infrequently
It's either bills or junk mail
I've cancelled cable, I've let the phone go
I still get the morning paper
It piles up
The stories never change

It's like I read the same paper every day
In fact, I have a favourite paper
Tuesday, May 6th, 1997
I read it every day after I scan the new one
and put it in the pile
In my Tuesday paper, there's a good story
about communication between animals
How to understand what animals say
I lie on my couch
My fever seems to have lifted
I stare at my white walls
The light coming in through my window
is stirred by the shadows of that locust tree
It's spring and the new leaves blow in the wind
We've had a late snow
But it's melting
Dripping on my windowsills upstairs
A kind of percussive thaw
When the weather is good, I feel like going outside again
I'll walk close to the buildings, then move out into the open
I'll go down by the river and watch the birds swarming
I'll go out and feel the sun and the wind
And be among the people again
The people who will never know me

I'm my own person, I'm one of a kind, I walk my own walk
I'm unique, discrete, disconnected, and hard to get a hold of
I'm surrounded by robots
They're just going through the motions
They listen to the same music and they recite the same lines
They have the same goals, they appreciate the same things
They're normal, they're cut from the same mould

Even their laughs sound the same
They're like ants or bees, they seem to be part of a big picture
that I don't belong to
I'm out of sync
I don't have a role and for that I'm thankful
because the direction they're moving in
is going nowhere fast or slow —
They're going nowhere, nowhere at all

It started with my parents, their expectations
They wanted me to be this or that
Actually, they didn't seem to know what they wanted
They just wanted me to cooperate
I couldn't bring myself to give in
It wasn't that they were so demanding
It just added up over all that time we spent together
They sent me to school

When the weather is good for a couple of days
I want to go out again — to walk alongside the traffic
to feel part of the whole
To watch the black birds swarm
to see the clouds drift across the sky
to feel the strong pull of the river
to feel the heat of the sun and the wind on my face

I watch the people wandering around, trying to manage all the
different kinds of influence, their imperfect relationships
they're trading partners or changing religions
flipping hairstyles or clothing or even changing
their facial features
losing weight, changing their diet

shifting their confinement or restrictions
drowning in the flow of the changing channels
But I've taken a different tack, I've learned how to reject
orders, to ignore demands
I still want to belong, but I feel more comfortable
when I maintain my distance
I don't mind being pushed around
as long as I can walk away
As long as I can maintain
my distance

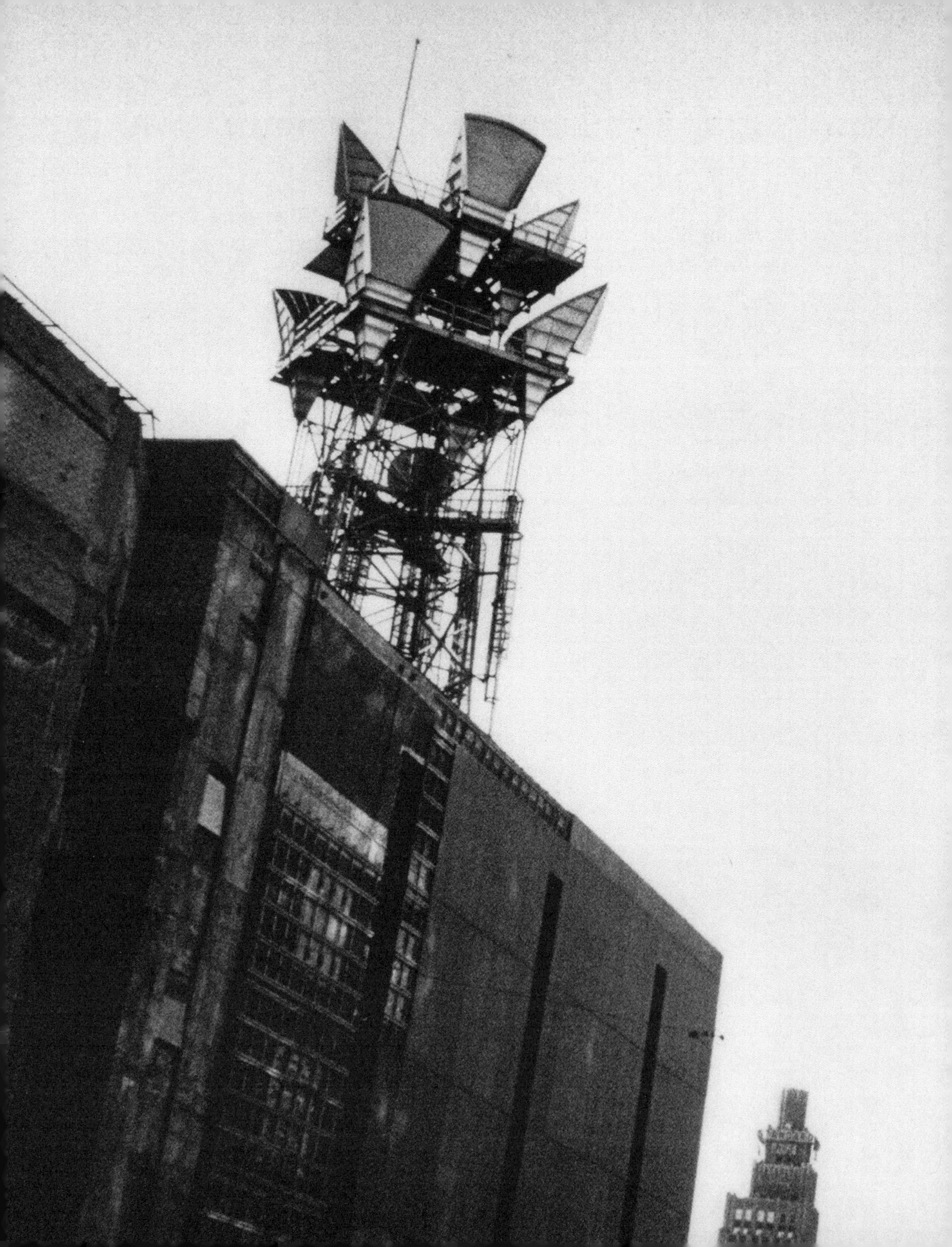

THE FUTURE IS HISTORY

TOM SHERMAN

Television is the History of Video 1993

[1] TELEVISION IS THE HISTORY OF VIDEO.

What does this statement mean? Maybe it would be clearer if it were phrased differently? The history of video is television. Maybe the problem is more serious than proper phrasing? The problem is that video doesn't seem to have any history. Video is something new on television. Television on the other hand is nothing new. Television is middle-aged. It's been around for more than fifty years. It hasn't changed much. Sometimes, it's live. Most of the time, it's canned. Whether live or prerecorded, the people who make up the face of television are mostly middle-aged. In fact, people on television are middle-aged, whether they appear to be young or old — because middle-aged people write all the stories on television. Video is different. Video is for younger people. The images and voices of video are those of younger people. Of course, there's video for the middle-aged. And everyone will grow older, and someday video will be nothing new. Video will be the history of something else. I wonder what that will be? Video in the future will be live and not just canned. Video will be picture-phones, and thus video will be everywhere. It won't just be on television, or played on a VCR or a computer. It will be live and it will be canned and it will be big and it will be small. It will be image and it will be sound, and it will be everywhere and nowhere at the same time. Because unlike television, where everyone used to watch the same programs in small groups or alone, video will isolate still further — with video, no one will watch the same programs. It's not that video was designed to isolate people. But video is the future of television, and in that tradition, its ultimate effect will be antisocial. Video as a medium will age and in time its viewers will become middle-aged and elderly. Young people will always want to watch something else. We will just have to wait and see what the future of video will be.

[2] The Future of Video Is Something That Will Produce Frameless Images.

The next big image machine will let the pictures out of the box. No more picture frames ... no more movie curtains ... no plastic boxes around those screens ... no restrictions on the size of a projected image. Imagine a machine that makes pictures without edges — an image-making or image-taking machine that functions like the human brain. The biocomputer makes pictures of our world without edges. We see the world whole. In our mind's eye, we see frameless images all the time. But we can't make them, yet. Our paintings and photographs and video screens have frames that separate them from our imaginations and our sensorial brains. This reality serves to continually revamp the power of objects. Found objects or sculptures have no frame. A ceramic vessel is the perfect gift in that it need not be framed. Objects have edges, but their image is not limited by frame. Objects are limited by gravity. What we need are images without frames that weigh nothing. We need images that defy gravity and escape the prison of the frame. Will someone out there invent a machine that produces frameless, weightless images? I'm not talking about holography or today's so-called virtual reality technologies. I'm talking about an image machine that can fool us. An image machine that makes or takes what appears to be reality in its actuality — and stores it and multiplies it and magnifies it. With the mental construction of this new machine, I'm talking about the future of video.

[3] The Next Major Image Technology Will Literally Install the Viewer within a Field of Information.

Looking through a window won't satisfy every desire. Some of us will want to be right inside. The observer wants to be part of what is seen. The viewer of a scene wants to be central to a field of visual information. Smack dab in the middle. From an absolutely central point of view, I spin around to gather information. I want someone to give me an image-making-and-taking machine that will deliver me to the centre of an image with no edges, with no frame. I want to enjoy full integration with the image of my choice. I want the objects around me to be real as I position myself inside an interior with full, rich detail or an exterior locale with horizon line circling, circling, circling around me. I want to be integral to the creation of reality. I can't get excited about telepresence or virtual experience. I want to really be part and parcel of what I see. Illusion is boring unless it is flawless. Install me flawlessly within an image of reality. Only then will I surrender my believing eye.

[4] Video Is a Memory Technology.

Memory has replaced logic and reason as the basis of truth. Memory is a place in the distance where we observe ourselves. We leave our bodies for the space and time of our minds. Images formed by a silent shower of light. The body suspended, the mind fastened tight. Memory is hard to control. Memory is recollection, and recollection is the trigger of imagination. We view the past as we must, in the best possible light. We second-guess the pain. Our minds are full of conclusions. The past is the truth in the shadow of doubt. The technologies of memory are often utilized as instruments of deceit. Memory is written and rewritten over and over again. Documentation is the sticky residue that binds together the loose ends of our imperfect memories. To verify all events within the broad context of reality, we must check and recheck the record. Our relationships with machines are documented by the recordings left behind. The shelf life of a video recording depends to a greater extent on our need to remember than on the physical conditions governing the security of storage. The archivist must save everything if he or she wishes to be remembered at all. The video machine has become our impersonal, inanimate archivist. Our intimate relationships with these machines are well documented.

[5] Video Is Antitelevision.

Whether it's called video by independent producers or video art or just plain video, it's really antitelevision. Antitelevision is made by individuals or small groups that oppose the psychosocial conformity television dictates to its viewing public. Television, with its ubiquitous presence in every home, defines our concepts of individuality and family and community and nationality. TV describes us as a species. If we actively oppose TV's governance of our minds by making our own video, we are producing antitelevision. Strangely enough, governments are known to support the producers of antitelevision through grants of public money. Tax money is selectively turned over to all sorts of minorities so they can represent and publicize their interests in video. An incredible rainbow of underrepresented types end up making antitelevision, which looks different from, or rather looks at different things than regular television. Unfortunately, antitelevision has little effect on regular television because TV isn't easy to push around. Television is anchored by millions and millions of people who watch it and the billions and billions of dollars spent to lure and rivet its viewing audience to the insidious glow of the tube. In fact, television isn't a technology or a network or a global confluence of electronic delivery systems. Television is essentially its own viewing public — the TV audience itself. Antitelevision is opposed to the tastes and mindsets of the regular television audience. No wonder we see so little antitelevision on television.

[6] I've Got 537 Channels of Television Coming into My Home via Cable.

I pay monthly bills based on the number of minutes I watch. It takes me twenty minutes just to surf across the surface of all my channels. I can read almost every channel in a second or two, but when there's something I can't figure out, I find the image arresting. I love uncertainty. It is such a rare commodity these days. Formula fiction, formula news. Information that fits the mould. Familiar faces. That certain sameness. Makes me a channel surfer. I'm riding waves of cop shows and nature shows, the talk shows and the movies, the talking heads, the dancing bodies, and the news, weather, and sports. And the news, weather, and sports. There are the standard international stories. There's sex. There's violence. There's war. There are games. The animals are simply going about their business. Tight young bodies find themselves connected to a universe of commercial concerns. Television delivers people. My cable delivers an ocean of television with its unnatural colours and its primitive sounds. Oooh. Aaah. Oooh. Aaah. Sit back, relax, the pattern has motion; a nice pace. Your pace. You are in control of what you watch. You can read images at a rate that makes your head spin. You've got over five hundred channels. You can read all of them in just under twenty minutes. You are in command. You want to be alone so you can concentrate. I understand. You just want a few precious seconds of every single channel your money can buy.

[7] Museums Will Continue to Thrive and Multiply Because They Will Provide a Public Network of Technology-Free Zones.

Technologies like video and computers are kept out of museums in an effort to prevent the collapse of the economy supporting the precious object. Museums inherited their unique position as the last technology-free zones when the various churches of the world disintegrated into a full spectrum of television broadcasts. But even museums, which appear to be completely devoid of contemporary technologies, employ surveillance technologies, which are hidden. Precious objects must be protected and those who view these precious objects must be carefully studied. If an audience is understood, it can be cultivated. The art-museum audience will continue to be monitored for purposes of collection security and psycho-economic research, but all surveillance technology must be completely concealed. Live security personnel will also operate undercover, dressing and acting like museum visitors, so as to alleviate stress incurred by the audience while they view precious objects under constant surveillance. Nothing beats the feeling of apparently being alone with a precious object.

[8] At the Close of the Twentieth Century the Tremendous Growth and Influence of Video and Computer Technologies Have Given Rise to a Resistance by Artists within the Video and Computer Culture.

Artists seek to reform these technologies through direct use. They seek to change the systems from the inside. Today's political, social, and economic institutions hold power largely because of the technologies they employ; technologies like video and computers. These technologies themselves can be seen as powerful institutions. Supranational machine-states. When we think globally, we are contemplating the magnificent, virtual potential of these machine-states. Global technologies in the service of people, of course. When people are shut out of these systems, it forces them to see these potentially vast technological domains in strictly regional terms. For example, a relatively inexpensive video format and an equally modest distribution system may become the local, somewhat crowded, domain of several competing communities of interest. Competition over technological resources shuts out a lot of people. On the bright side, these technologies are dropping in price as they rapidly link together into networks and fuse into a comprehensive global net. Human nature is superimposed over the planet as a brilliant field of electronic memory equidistant from our individual physical bodies. All aspects of machine reality must be experienced remotely. Out-of-body experience has never been easier to achieve.

[9] In the Information Age Even Entertainment Has to Be Informative.

A lot of people just can't enjoy themselves unless they feel that they are learning something. That's the reason fiction seems to be so tied up with real life these days. It's gotten to the point that even our fantasies are documentaries. They are wiring up the dreamers and counting erections and twitches. Six and 234. Four and 116. Who are *they*? The social scientists, analysts, researchers, and marketing experts. But it doesn't matter how we perform or what they expect from us because most of us don't produce information. We consume it. And in doing so, we destroy it. We make a mess of everything they put before us. We suck everything they offer into our messy little moist brains, and we store some and we ignore more and mix things up to the greatest extent imaginable. Deep inside. In the dark silence beneath our voice. We make it happen. We put things together. We make our own information. Information is concocted every day in pretty average minds. We create order out of chaos. For our survival, for our amusement. There are professional information providers, the code makers, the programmers. These professionals shove the symbols down our throats in the image of entertainment. The only thing we can say about them is they are consistent. Redundant even. The sheer repetition and sameness of their engaging formulas proves they reveal nothing but the absolute truth.

[10] When and How Does Data Become Information?

Before supper, I asked to be excused so I could take a walk out across the field behind Norman's house in Prince Edward Island. I needed to establish a rapport with the red clay landscape on my own. A late-afternoon sun charged the vibrant reds and greens of the rich, rolling earth as I set out on what I expected would be a relaxing walk in the country before supper. I crossed a plowed field where the red clay was dry and dusty. My feet were stirring up little pink clouds. I stepped into an immense field of high, yellowed grass where the wind blew steady and strong from the west, playing a familiar song in the trees that lined the valley. I was thinking about my work back in Halifax, my love in Toronto, my life in general. I was looking into the horizon, but I was seeing myself. I was walking along at my own pace, basically unconsciously. When suddenly there was an explosion of motion at my feet as a rabbit bolted out of its hiding place, zigzagging across the field, disappearing in seconds into the high grass ahead. Startled, I scanned the open field with wide eyes for more wildlife. It is simply amazing how quickly the taste of adrenalin comes to the mouth. My heartbeat had quickened, my breathing had stopped, my level of attention had shot up to the degree where I was on edge. At this point, I began to see the landscape with great clarity. I noticed several places where the brush thickened, and there was one dark spot just ahead that looked like a hole. I approached this hole with some caution, realizing a few steps from the opening that it was too small to be the entrance to a rabbit's den. There I stood, intensely focused into the deep, dark hole. Just wide enough to force my hand in, but too deep to reach to the end, it could be a snake hole. I could almost see a big, black shiny king snake emerging ever so slowly when I felt a sharp pain in my wrist. I slapped dead a bloody deerfly, smashing it where I usually wear my watch. The damned fly had been feeding on me, and where there is one biting fly, there are always more. I had to get moving. I started walking briskly. I was excited, my blood pressure was rising. The wind was raw in my ear canal. I felt the heat on the surface of my skin. Was it the wind or the sun that made my skin burn? I had had enough of the country. I headed back for the shelter of the house. It was at Norman's house that I wrote "The Rabbit Theory of Data Transformation."

[11] To Be of Value in the Information Age a Work of Art Must Be Loaded with Information.

To Be of Value in the Information Age a Work of Art Must Be Devoid of Information. To Be of Value in the Information Age a Work of Art Must Transform Data into Information.

TOM SHERMAN

Reflecting on the Future of Video 1991

It's Cold at Night

It's after midnight on a cold winter's night in Ottawa, toward the end of 1991. The apartment is quiet except for traffic noise, which drifts in from a distance through the open window in my studio. Both my loved ones cough periodically from their respective bedrooms. A lingering flu-bug takes away from their rest. I have a mirror in my studio and find some humour in watching myself grapple with the ideas in this developing text.

I stare at a video-score pinned on the wall above my desk between these thoughts. I've always blocked out the timing of tapes in drawings and diagrams before shooting and editing. The more video comes to resemble music in terms of composition and movement, the more sense it makes to score it like music. I've been at this video business for over twenty years now. Things have changed since the early seventies, and yet nothing has changed at all. Video as an art medium still has a tremendous, largely unrealized potential. Despite its continuing shortcomings, video is still the medium of choice for many artists. There are more artists working with video today. There are hundreds, maybe over a thousand now. Tens of thousands of hours of video art have been recorded since we started using video to make art. We've made a valiant, concerted effort to lay claim to a technological domain that has expanded beyond our wildest dreams. Video, the technology once rare, is now common. But the audience for video art hasn't grown dramatically. Video artists and their limited audience still make up a small aspect of today's expanding video environment.

Video art is a small business sector when compared with the broader field of music video, instructional video, real estate video, and all those things now called video — video movies and laser disks, home video, desktop video, computer games, and so on. If anything, video art makes up a smaller share of the total video environment than in the days when the technology was truly a novelty. In fact, in the early seventies, when video art was a strange anomaly within and in parallel to the postsecondary educational system, artists seemed to be the only professionals who could make the medium live up to its then severely limited potential.

I think of video as a technology, as an art form, and as an environment in which I have worked and will continue to work. I am an optimist, believing that video will continue to offer a more and more enlightened machine-enhanced perspective of the world we live in. I am frustrated by the current shortcomings of today's video reality while admitting that I am continually intoxicated by the medium's fantastic potential power. I have bonded with a particular machine for life and wish to continue to develop within this relationship.

Video Is a Machine

Video as a technology is always changing, advancing. Professional technologists, the engineers, perceptual-behavioural psychologists, market analysts, and designers determine the rate of change and the nature (physical characteristics) of the machine itself. These professionals are paid to develop the technology of video by the shareholders of multinational corporations. Profit motivates machine development. Machines can be developed only as quickly as the market will bear. The consumers have to adapt to the changing nature of the machines. Relationships have to be continuously upgraded. If the relationships between people and machines become meaningless and unsatisfying, there will be psychological and economic dysfunction.

Artists are professional developers of the technologies they employ. As artists must gain access to video technology to make video, we become the quintessential "prosumers." Artists team up with engineers, perceptual-behavioural psychologists, market analysts, and designers as professional technology developers. Artists find themselves in the ranks of those serving the interests of the stockholders of multinational corporations, continuing to develop the technologies for tomorrow's prosumers. As perceptual and social "scientists" we employ ourselves not for the profit of the shareholders of the multinationals, but toward our own gains. Although the contribution by artists is equal to, if not greater than that of other types of professionals on salary, we are not financially rewarded directly by the corporations. This is so because we insist on determining our own set of goals, methodologies of production, and pace of development. In some cases, we are way out ahead (artists are unrealistic in the demands they put on the technology), and sometimes we insist on applying the brakes (artists respond in a reactionary manner, insisting that the technology be used only as it has been in the past).

Today, video is a ubiquitous, arguably transparent technology. It is commonplace and offers the individual the virtual potential to be everywhere. Although this virtual potential is alluring, offering a dream of electronic dissemination far exceeding reality — however unrealistic — video does encourage us to cross disciplinary boundaries, to work with others, to expand beyond ourselves.

The video machines themselves are becoming more sophisticated. Today's machines are more demanding within our person-machine relationships. We all form relationships with the machines we use for "reading and authoring" the world. It is possible to develop a social psychology of person-machine relationships. Our relationships with our machines are changing. On the one hand, they are easier to use. But as video and computers and laser technologies merge in the digital age, we are forced to expand our perception of context. I here define *context* simply as the territory within which we must work. Once we were able to think of our territory as a discrete system of video machines connecting us to an audience. This area was the only real territory we had to work with. Video as a work site is far more extensive today. The video domain now includes aspects of the media environment that video artists once considered peripheral to their video-biased scope of vision. In the strict definition of hardware, video is becoming peripheral to the digital computer. Video is now commonly defined as the input (camera) and output (monitor) peripherals of the digital processor. One day, everyone's relationship with video will be a relationship with a video-computer. The idea of video as a peripheral is now part of a relationship artists share with the intelligent artifact itself, that external representation of our intelligence, the artificially intelligent computer. I sometimes think we live in a vast domain of integrated communications systems tied together in homage to the concept of artificial intelligence. As the territory has become far more extensive and complex, we find that we have been marginalized as specialized artists who capture images with a video camera.

It's Never Too Late to Change

Change is constant with all communications technologies. These technologies are never fully developed. There are no finished products. Technological development is a process driven by those who design and those who use the gear. Video is an information machine, a communications device, an extension of our bodies, our societies, and our reality. Video information is transformed by all those who use it. Video is worthless unless it is exchanged or, more importantly, used. Video is a manifestation of electric speed and power. Video is a vast source of memory. Video is seductive on both sides of the chip. The cameras and monitors and the processors in-between are instruments for transforming the world within reach of our eyes and ears.

Access, Access, Access

Before I stroke the keys more toward a description of our common video domain, let me return to the predicament of artists working with video in the late twentieth century. Video artists face a set of problems universal to all contemporary artists, regardless of the media and technologies they choose to employ. These challenges can be summarized as the perpetual problems of access, problems that will dominate our future. Access means the availability of anything that will allow us to achieve our goals. Video artists need access to the time and financial resources necessary to make their art. Video artists need access to the technology itself. Video artists, like all other artists, need access to an audience. Access to these necessities must be earned and secured through a sustained intellectual, emotional, and political struggle.

The first challenge for any artist is to find a way to exist as an artist. How does one find a way to make a living without compromising one's ability to make art? This challenge is, of course, a universal problem that remains fundamental to those having chosen the profession. Some way must be found to have the time to think clearly, to observe, to question the society at large. At issue is the quality of life itself. The rent or mortgage must be met, food must be put on the table, bills paid, means of transportation and communications secured, a community formed, friendships made and maintained, vices controlled, desires satisfied, and love found.

Psychoeconomic factors at the heart of the artist's day-to-day existence are even more important than cash economics. I use the notion of psychoeconomics to introduce concerns for quality of life from the perspective of the artist. Psychoeconomics is the study of how economic conditions affect the psychological state of the individual; it goes way beyond whether or not the artist is worried about money or issues of class. Psychoeconomics are also based on the individual's ability to give and the capacity to be influenced. Artists have traditionally valued their independence above anything else. Independence is measured in degrees of freedom balanced against acceptable levels of responsibility and the absolute necessities of compromise. Artists are generous only when they have achieved a certain measure of security, which is attainable only if they have received recognition from those they respect.

If we consider these general indicators of emotional and intellectual well-being within the context of a specific professional community, such as video artists, it is possible to predict a conformist behaviour in the video-art field. Self-esteem depends on maintaining a presence in the aspect of the media environment that guarantees recognition by a respected community of interest. Artists tend to cluster in peer groups with other artists who share the same values. Other related professionals — administrators, critics, curators, academics, programmers, distributors — flesh out these communities of interest, all maintaining their own professional alliances. The video-art community is an international community of interest that functions like any other professional community with shared goals. Individuals and small companies are linked through institutions, events, and documentation. Recorded video information is exchanged through the mechanisms of personal correspondence, video publications, screenings, exhibitions, festivals, cablecasts, and broadcasts. There are professional newsletters, magazines, and books to substantiate video information exchange in print and still photography.

Poor Video Artists

The economic reality of the video-art business has always been and remains grim. Competition is fierce — the rewards are meagre. Video art is an information commodity stuck in no man's land as a perishable good with marginal distribution. Although the new economists of the information age sing the praises of creativity, there are virtually no public or private resources forthcoming to cultivate demand for video art. Thus the audience is nurtured at a snail's pace. Video art rots with its tape stock on the shelves of selected archives. Video-art history is no different from other studies of our disposable image culture. Like film, television, sound recordings, and radio, the study of such perishable information goods is usually conducted within the stable confines of print, supplemented by still photography. Today's computer databases list titles and general descriptions of what amount to a skeletal, text-based index of productions. Organizing video information by descriptive text index, while obviously important, is but the first step.

Commercial media sectors, such as the music recording industry, have the means to revitalize and exploit their artists' past accomplishments. But the history of video art must still be cryptically perceived through oblique references offered by this year's tapes. An astute, informed viewer may pick up references in new tapes to previous video works. In fact, video art is a series of statements by "authors" aware of the "literature" of the field. In the absence of serious written criticism on video art, artists in the field tend to make critical, aggressive tapes designed to challenge positions established by previous video works. Video by artists is not designed to serve an audience the way commercial or educational media are. Rather than give audiences what they want or what they need, the artist offers a personal communiqué, a statement of individual concerns, beliefs — a kind of declaration of independence. The cost of this independence is high. Even today, in a time widely perceived as the video age, an artist can't make a living based solely on the sale of his or her video art.

The video-art business operates on the principle of pure information exchange. The exchange of video art in 1991 constitutes an advanced information micro-economy. Video recordings are exchanged between individuals, small companies, and institutions, primarily in video-for-audience transactions. There are small financial transactions tied to these video-for-audience exchanges, but to date the modest cash income tied to video exchange disappears quickly as it is applied against far greater expenses of production and distribution. The video-art commodity is typically financed by the artist (the entrepreneur) through government grants and/or the investment of major equipment manufacturers (most commonly the temporary loan of equipment). No matter what the perspective of analysis, financial exploitation of video by artists, distributors, and exhibitors has clearly been minimal to date, considering the volume of video information exchanged for audience over the first twenty years of medium's history.

Degrees of Influence

The rewards for the primary information providers are measured largely in terms of effect. Artists intend to effect change through their video works. Return on investment is measured in terms of influence. The attractiveness of the medium is essentially its power of influence. Artists strive for a presence in the video environment, hoping to shift the perceptual or conceptual base of the audience or to contribute to a transformation of psychosocietal norms. The individual who chooses to negotiate reality through the production of video is generally someone who wants to effect change. The medium (video used by individuals or groups of individuals to effect change) is not well suited for engineering social stability. In this respect, video is the present and future antithesis of television. Video will clearly emerge as the third aspect of our predominant media triumvirate: cinema-TV-video. Television will continue to rule, with cinema as its past and video as its future. Video will transform television over time. Video will continue to be distinguished from television by individuals until the medium is perceived to be the quintessential representative form of the individual. Video will most likely share audience with the novel, short story, and literary essay in the future. Today's video by artists allows a community interested in change to anticipate and preview the way the individual will negotiate reality in the future.

Given this outline of the psychoeconomic factors affecting the video artist, it is possible to project general expectations of behaviour for the near future. Artists will continue to use video as a catalyst for change. The medium's proven effectiveness in presenting alternative models of sociopolitical theory and practice will not be overlooked in the years to come. The video documentary, including a variety of "docudrama" forms, will continue to interest artists obsessed with reforming their sociopolitical reality. Let's call these kinds of artists social reformists. Their video will continue to oppose the predominant ideology of television, though a legitimate concern for message clarity and accessibility will result in programming virtually indistinguishable from commercial and educational broadcast media. The social reformists will attempt, through the use of appropriated video formats, to propagate different values to an audience considered to be poorly served by commercial and educational television. Thus the social reformists will continue to build minority cultures from scratch, using video to establish communities of interest in an era of social fragmentation and alienation. Other kinds of video artists — let's call them the individualists — will concurrently persevere in their efforts to distinguish themselves as individuals against the orthodoxy of individualism portrayed in the dominate forms of commercial television and cinema. Although this personal activity will be perceived as being counterproductive by the majority of video artists, namely the social reformists, it is the elemental breeding ground of fresh, new programming trends. The only way to capture larger audiences without industrial-scale financing is through creativity. Literature (writing and publishing) provides us with a hint of this opportunity today. Small literary presses, journals and magazines, although threatened by rising costs, point to a necessary future development in the video domain. Video publishing will inevitably replace the small literary press.

A History of Hands-on Relationships

None of this expected behaviour will occur if artists fail to continue to secure access to video technology. If we consider the history of the medium, access to equipment is less of a problem today than it was decades ago. Artists have generally had to work with equipment on the bottom end of commercially available technology. When I started making video, the bottom end was 1/2" reel-to-reel. State-of-the-art equipment was 2" broadcast gear. The technical discrepancies between these formats were absolutely debilitating. Video dubbing or editing across formats was rarely technically possible. Video recordings played back on someone else's "compatible" machine would rarely provide a stable image. There were no reliable edit-controllers in those days. Colour 1/2" video wasn't available until 1973, after six years of black-and-white instability. Playback decks and decent monitors were scarce. Conditions improved gradually over the 1970s. The bottom-end compromise to high-priced Betacam and 1" professional gear became 3/4" video. The relatively recent development of 1/2" consumer VCRs and camcorders and, more recently, the dramatic emergence of 8 mm consumer video [and hi8 and mini-DV since then] have given a lot of us reason to be optimistic about the future. The technological infrastructure of video has developed far more quickly than many of us originally imagined. Today, most artists are still working on the bottom end. So everything has changed, and nothing has changed. But today's bottom end is like the Cadillac of the late 1960s. Our tapes can now be played on literally millions of playback systems worldwide, and most of the time the replayed recordings are true to master quality. Image and sound quality is still limited by production and post-production budgets, but we are no longer strangled at every juncture by erratic machines. Government grants are used to buy our way into the future: everybody seems to be "bumping up" recordings or making special efforts to climb the ladder to technical formats on the top end. Of course, today's high tech is tomorrow's bottom end. The ladder is a treadmill. We have survived to see desktop video publishing. Personal computers now process video signals. Video images can be "written" in software. The video camera can be and is considered by many to be a peripheral device. Our computer-video signals can be stored on disk or tape and played back on our neighbour's televisions with stereo sound. Our camcorders are now truly functional in low light, and they are physically lightweight enough to be comfortably hand-held for shooting. Artists will always want a better image with the most natural sound. We want digital video and know it is just around the corner. We want more precise control over our video image and audio. We want to be able to reproduce our tapes without amplifying or introducing noise. We want access to the best that money can buy, and we want to use the prototypes, too. In the future, we will find a way to get our hands on the best technology available today. By the time video artists satisfy their current demands, today's best gear will be tomorrow's bottom end.

Be Prepared to Stay in School Forever

All of this "progress" forces the artist to upgrade his or her skills continuously. Continuous education replaces continuing, periodic reschooling. The machines advance into states of increasing complexity and power, committing those who follow to an exhausting cycle of skill development. Those who choose to run along for a full revolution always find themselves back where they started. This is the machine's most basic philosophy course (10101010101). Progress always has its downside, and the conclusion of any race is a return to the starting line. The digital revolution will certainly be interesting to watch. Analog wave forms will be cut into digital bits only to be reassembled analogically again. I'm sure there will be nursery rhymes written about such remarkable transformations. Our children will write them for their children's children. Tomorrow's digital children will continue to sing new versions with imperfect voices, until their throats are upgraded. Learning is easy when you are young. Running along with the new technologies will bring us full circle. In my youthful muse, I remain optimistic.

The final challenge is the greatest access problem, that of audience. The access problem of problems is the challenge of finding an audience. If the video artist could reach the potential audience for his or her medium, access to time, resources, and technology would take care of itself. Unfortunately, reaching an audience of any scale has been hugely problematic. The ultimate frustration of working with an unlimited multiple form like video is to find one's recorded information decaying on the shelf unused. But then again, most video works by individualists appear to be designed to function as anti-information. I am thinking along the lines of a mixed analogy, comparing underground video with ideas put forth by nuclear theorists (that is anti-matter). If we consider the general opposition by video artists to television, the mass circulation of video information in standard programming forms, it is possible to assign value to static, apparently dysfunctional, video-information forms. Individualist video art is particularly convincing as anti-information when compared with the unrelenting conformity of television. Given the generally negative view of television expressed by its regular viewers (most think TV is a waste of time), how long can video anti-information remain so ridiculously undervalued? If government would only tell people that watching too much TV is a bad thing, that it will make them sick. And that watching video (antitelevision) may offer a road to recovery.

It is possible to imagine a video-information dead-zone (dead only for the time being — a repository of antitelevision) stocked by society's forgotten fringe celebrating the formal, however subconscious, awareness of their death wish. The present and projected capacity of this video dead-zone would seem to be unlimited, given the current modest rate of input by video artists. This negative concept of zone capacity might be as abstract (and as useless) as an inconceivably immense field of unaddressed memory.

What Audience?

I would like, for somewhat perverse reasons, to introduce the audience for video at this point. What is needed is an audience that can explore the depths of the video dead-zone. In these days of early exploration and development, the organizers and cataloguers will have to suffice as the video artists' primary audience. These primary, pioneering researchers will quietly mine the resource and provide their findings to the institutions they serve. The role of the researcher is to recover through excavation. Institutions then frame the specimens and diffuse their value to their respective audiences.

The principal researchers at this time are the artists, curators, and archivists who follow a natural drive to know and associate with the work of contemporary artists. In all three of the above professions, we find the majority conducting their research with at least one foot anchored inside an institution. Artists in their artist-run centres, curators in their galleries or museums, and archivists in the archives or libraries are generally employed to conduct research and to organize and publish their findings for the benefit of their respective audiences. Unfortunately, the bureaucracies of any of these institutions, including the publishers and the cablecasters-broadcasters, are preoccupied with maintaining and increasing the size of their established audiences.

Any time anyone introduces new information to an audience there is going to be resistance. Compounding this natural conservatism is the reactionary stance of the critics. The critic serves the audience by reinforcing its fear and rejection of the new forms of art-information. Criticism has its own script, and its authors are intolerant of activity that doesn't fit the mould. Recognized artists should be consistent, and aberrant gestures by both young and mature artists will be ignored. Video made by individuals for individuals presents new problems for audiences attracted by other means. There are standards of presentation to be upheld, and there are programming demands made every day by audiences already established and served. Video art is designed to accelerate psychological and social change. The medium is a major information platform for presenting opposing views. Its producers, as in information producers, make anti-information for a subculture, call it the art world or the community served by cultural workers, and also seek recognition somewhat paradoxically from institutions that have established audiences through other strategies. The status an institution offers is in fact determined by its audience. Established audiences want tolerable variations of the kind of programming that attracted them in the first place. The notions of status advanced by the establishment must be transformed and new institutions must be struck. Meanwhile, video hardware continues to advance as an indispensable, cost-effective audiovisual peripheral technology. The equipment inventories in all manner of institutions are swelling. The new buildings are laid on a foundation of fibre optic cables linking an architecture of future viewing areas.

Anti-Information for the Antisocial

The largest audience doesn't go out at all. There are more and more display-playback machines in people's homes. Networking by computer is raging on, unregulated. The audience for television, video, and computer services prefers to stay home. Who is going in after that audience? First, everyone who can afford to spend a buck to make a buck is going to try to enlist the home audience first. The next people in will be those obsessed with influence itself. The peddlers of pure influence, those not suited to mainstream politics, will have to wait for the information utilities to develop before prices come down to permit entry at their level of financing. In the meantime, they all want the audience to visit their place, to hear their music, to read their books, to feel the power of their machines. Everyone wants some attention and a chance to influence someone. Video artists compete with everyone for audience.

The video dead-zone itself has an established audience of sorts, archivists and librarians, though it's hard to imagine an "audience" that is less opportunistic. Of course, it's difficult for archivists and librarians to find the time and resources necessary to exploit the vitality of the public and private holdings they manage. Shelf life is hardly a dynamic life force until it is tapped. Unfortunately, forgotten video programs are not going to go out to find a new audience by themselves. A new audience can breathe life into something off the shelf only if it has been given guidance on how to apply minds to the task. But when will a major home audience, like the television audience, move in the direction of video art?

Future demand for video information of all kinds is a given. Consumer and professional investment in hardware is growing at phenomenal rates and will slow only if the human race vanishes from the face of the earth. The technological infrastructure is developing relentlessly, inevitably. Electrical power continuously supports strict routines of exposure to an awesome stream of signals, symbols, and noise. We watch and rewatch. We drink the image-brew that is our daily flow of information. Media information received within the physical context of our living space has virtually become the sum total of our reality. Finding those of common mind over distance is our life's work. Controlling how we are represented in media is essential to our survival. There are others like you out there. Video is an appropriate technology for this communications mission. The video-computer is an even more effective device. Digital-video storage and the power of digital-indexing techniques (for assigning the act of composition over to the viewer through interactive systems) will certainly prepare us for our initial contact with the larger audience. The new viewers will be demanding. How will we meet their demands after we have managed to attract their attention?

Given the radical evolution of the technological environment, it is possible to foresee a time where real changes take place — where the potential of technological innovation (opportunities for new behaviour) overwhelms the natural human trait of apathy and the species' general tendency to resist anything introduced by strangers. In the late 1990s, video art (diverse programming produced by individuals who strive for change) will finally be exploited commercially by middlemen and middlewomen to a significant degree. The home audience, by that time, will be fragmented into minute, personal internalized communities of selves (such complex individuals can call upon a range of personalities to match up with different viewpoints and perspectives in programming). The elderly, if they have managed to maintain sharp memories, are a marvelous potential audience, because they have experienced more and have the real sense that they have been different people in the past. They may feel they have been several different people over the years. Although I recognize the tendency in older people to be who they are and nobody else, I do believe they can be put in a position where they will be stimulated again by the rich psychological histories of their lives. If I am right, there will be strong demand for works that elicit multiple perspectives, formed over longer and longer stretches of time.

The Formula for Success

My instincts tell me to keep producing video. I need to secure my own address on the information platform of tomorrow. The audience will come around when they can relate to my message. I have been patient. There is too limited a demand for anti-information today, but the audience's receptivity to change is increasing. The purveyors of redundant, standard forms are unwittingly creating an appetite for alternatives. The relentless proliferation of machines is also fuelling this demand. Steady increases in channel capacity provide additional room for aberrant behaviour and ultimately encourage program diversity. If you sell people machines that promise the freedom to create and choose their own information, some of them are going to hold you to it.

In the short term, I'll shoot for the attention of aging and elderly women. I think older women are more open than men to different ways of thinking and seeing the world. I'm talking about women from fifty-six to seventy-eight. In this age group, the younger ones are still likely to be living with semiconscious men who are attached to a second TV and different programming (mostly sports) in another room. The older ones, if not in a seniors' home, generally live alone (having survived their mates) or with their children. Their children are already middle-aged. Their children are in what is called the early aging block. This middle-aged audience will watch more and more video as its members age and become less physically active.

Information will be as basic a need as food and shelter to this maturing audience. A good percentage of disposable income will be set aside for things to watch. An army of video producers will be needed to keep these people amused and signed up to information services for life. The same forms of programming will fail to maintain established audiences for the long haul. Programming must grow in complexity to serve a maturing audience. The concept of programming must shift to meet the audience's evolving characteristics and nature. All aging audiences will demand new ideas, images, and sounds as they rediscover their second youth up ahead, beyond their careers. These desires will be recognized by the industrial producers, which will move to commercially exploit the large, aging audience to the greatest possible degree. This aging populace will soon be longing for youthful ideas.

Forget the grandchildren — they seem to exist for the privacy of their rooms, where they work late at night on their interactive toys. The kids will choose to communicate with their own kind. Picture-phones await future generations under just about everyone's Christmas tree. How many aging communicators will be capable of mastering the picture-phone's operation? How many aging and elderly women will choose to actively manage the programming options for their family's or their own viewing? Where will they find the money to pay for their new taste in video? How many of these senior viewers will want to originate video of themselves to re-establish their existence in the world?

I'm optimistic about the future. There will be so many aging and elderly viewers by the turn of the century. They'll all be sitting around staring at a video display of some kind or another. Video will also be in most of the moving vehicles by that time, for those who insist on actually going somewhere while they watch. Both the aging and elderly will gladly sit still, to watch video images fired at them through a CRT (the television tube). Technological advances will continue to outstrip our abilities to be critical of changes. Chip cameras and digital-storage devices will become so small they will be implanted inconspicuously in a number of people's foreheads. Initially, these video implantations will be in aid of the handicapped and aged, and very useful to secret service agents and private investigators involved in corporate espionage.

I won't hazard a guess on the first year of widespread video implantation in humans. But video will continue to be known as the most immediate, intimate technology for some time to come. Video will continue to be used to soften people up for the computer. The whole human race will add video-computers as important components of their personal communications links with their neighbours and the world. Somewhere within this well-equipped populace, there will always be a minority supportive of those who advocate change. This minority will always be critical of programming serving the complacent majority (television). Video artists will continue to choose to serve this critical minority by providing video featuring alternative models of social behaviour and new perspectives for the complex individual.

Video's future, where will it be? The places for video will be in the home, in the theatre, and in its myriad installations — its unlimited attachments to the world. When you grow very old, you will become more and more attached to video because it will certainly be everywhere then, just as it is now. Eventually, video may even be attached to you.

Turning Television into Radio 1991

I've been working in video, against the forces of television, since 1970. I am an artist and writer of American origin, though I've spent most of my adult life in Canada, having emigrated to Toronto toward the end of the Vietnam War. It is from my Canadian perspective that I offer these thoughts on television and the way it affects my languages — the visual, written, and oral expression of my voice.

Although I have worked in many media (drawing, painting, sculpture, photography, text, computer graphics, robotics, video, film, audio, television, radio, magazines, and books), seeking appropriate forms in which to speak in various ways as an artist, video has been the medium that has served me best. I have literally grown up with video, as it has concurrently matured as a communications technology to its present ubiquitous middle ground between cinema and the computer's digital realm. For as long as I've had something to say, I have played video off against other media as a means of securing my presence in the world.

Video represents the devolution of television, in the sense that with video we find ourselves moving backward toward radio, with its immediacy and great variety of voices. As Marshall McLuhan pointed out, all previous media become the content and nature of any new medium, and the reverse is also true. This transformation is a natural and sensible process of cultural recovery and balance, as we race ahead into the next millennium. Nevertheless, the devolution of television must be recognized for its broad and revolutionary implications. With video, we see diversification of television content and the resurgence of the individual voice. Where television has overwhelmed and forged uniformity in the voice of the individual, video is rapidly turning television into radio, returning power to the individual. Not the radio of the past, with its one-way dissemination of scripted drama, news and recorded music, but the radio of today, with its open-ended talk shows and integration with the telephone.

Later in this text I will discuss two of my video works that embody this devolutionary theory. These are "Television's Human Nature" (1977), a twenty-eight-minute program to be broadcast or cablecast, and "Exclusive Memory" (1987), a three-hour video recording for installation in museums or other public viewing sites. These two works are essentially about television, using video to deliver narratives in natural, spoken language.

To begin, an outline of my background and an introduction to ideas of visual, written, and spoken language. I was born and raised in Manistee, Michigan, a small, isolated town in the Great Lakes region. Although not as central as Iowa or Nebraska, Michigan is part of middle-America, landlocked in the centre of the continent, culturally isolated from both New York and Los Angeles. I gathered my early culture from Milwaukee, Chicago, and Detroit regionally, listening to the latest pop music on early portable transistor radios and later on car radios, and expanded my horizons through long-range radio listening ("DXing"), using sophisticated communications receivers and elaborate long-wire antenna systems. Listening from my room late at night, I was able to monitor cultural trends on both coasts of the U.S. and throughout Canada on medium wave, as well as worldwide through the short-wave services of the BBC, Radio Canada International, Radio Australia, the U.S. Armed Forces Radio Network, Radio Moscow, Radio Havana, and so on. Radio was my means of overcoming isolation.

Television was extremely limited in those days. I'm talking about the 1950s and 1960s. I remember watching television for the first time in 1952. My dad brought home our first TV the summer I was five, just before I went to kindergarten. There was only one channel captured by "rabbit ears." The first program I ever saw was the *Captain Hal Show* out of Green Bay, Wisconsin. Captain Hal hosted his show from the wheel of his ship. He would invite kids to march around his deck and play silly games, cutting away to *The Little Rascals* or cartoons for most of his show. At the end of his program, he would ask his viewing audience to pick up their toys — to be good little girls and boys. To check up on us, he would look right out into our living rooms with his spyglass, calling out our names. Nearly every kid in America could be covered with six names in those days (Christie, Karen, Susie, Johnny, Jimmy, and Tommy), and I was awestruck when he called out my name. I was convinced that television offered Captain Hal the power to watch me viewing him. I went off to school for the first time that fall, believing that television was a fully interactive, two-way medium. I dared not speak to Captain Hal, but apparently the option was mine. I've used this pseudo-interactive technique in many of my video works, including "Television's Human Nature" and "Exclusive Memory," addressing the audience as if they were with me "live" at the moment, feeding off their immediate attention and demanding their internal response.

Later, I learned to watch television the way everyone else did, half asleep and basically inert. Our reception had improved with the installation of a fancy antenna on the roof controlled by an electric rotor. I watched the Westerns, *I Love Lucy*, and *The Untouchables* to escape reality and waste time. To supplement my interest in cultural trends, I watched *American Bandstand* and *The Ed Sullivan Show*. Parents weren't concerned with keeping their kids from watching too much television in those days. I can remember how I'd lie on the floor, watching everything that came on, until I was stiff as a board. I would get grumpy and would snap at my mother when she'd try to push me out of the house for some fresh air. She just wanted to watch her shows when I was outside. We were an average family. We watched TV as much as we could every day and fought over which shows we'd watch. We ate our meals in front of the tube, and bedtime was more or less determined by the *TV Guide*. I finally bought my own TV when I was a teenager with money I had earned from a summer job. With my own TV, I broke away from the tension of family viewing, retiring to my room to vegetate in the thick fog of my adolescence. Fortunately, the four channels available at the time (all fuzzy as I was back to rabbit ears in my room) signed off early enough, by 1 a.m., to permit me to escape into the world of radio. After the last national anthem had given way to snow on the tube, I'd get my second wind in the broad expanse of my radio dial.

Even then, a couple of major differences between radio and television were apparent to me. Radio was something you listened to as you did other things, but television was a completely dominating force. Whatever was on the screen demanded my full attention. I pretended to do my homework in front of the TV, but I was never able to elude its control. I later realized that only those younger than I, bonded with TV since infancy, could do other things while watching. In other words, television has never been background information for me, as it is for the present generation. When TV is on, I am drawn into the programming to the complete disruption or exclusion of other activities.

Unlike radio, television is not a trend-setting medium for spoken language. Our spoken language is in high contrast to the mundane speech of TV, with complex local eccentricities of vocabulary and syntax. Back home, we cultivated our local identity through elaborate, exaggerated verbal play. Although this play was partly a small-town ploy against the urban tyranny of media, we were far more willing to incorporate the inventiveness and style of radio personalities into our speech, however urbane, than the generic voices of TV. Radio was hip and television square.

It wasn't just that small-town America wasn't represented on TV. Nothing real was on TV. Television, even "live" television from small cities, always strived to be like TV from the big cities. TV language was bland, without accent and affectation, generic, demanding contradiction through colloquial speech itself. Hit music on the radio was universal — a kid in one town loved the same music that every other kid on the continent loved — but the things the deejays said between the tunes, the way they pronounced or lingered on particular words or ran humorously inappropriate phrases together, was what made radio real.

There wasn't any good television for young people in those days. Television was part of the adult world, something you wanted to stay up late to watch. Sure there were shows for young kids. There were always puppet shows and cartoons and cowboys and Indians and science fiction. But real things on TV for young people between the ages of eight and fourteen were just as scarce then as they are now. When the time comes to find yourself through language and social graces, television lets you down, but radio gives space to breathe.

It was the tape recorder that brought home to me the true power of communications media. With my first tape recorder, I could not only tape the music I loved or share the voices I admired with others, but also author my own voice messages and send them out. I had tried writing letters to kids I didn't know, but pen pals didn't excite me. Writing was too formal and dry. Then I discovered a "tapespondence" newsletter, associated with a monthly DXers' network bulletin. Others who listened to radio as much as I did, to expand their information-poor prison of a world, started exchanging reel-to-reel audiotapes through the mail. As a kid I could communicate across distance, sending and receiving recorded voice messages via parcel post. At one time, I probably had a dozen steady tape-pals. I'd hurry home from school to hear their tapes. We would all ham it up, issuing strange, exaggerated messages about our existence. We would start with the basics, the routine, and tragically mundane, and then begin to stretch the truth to make things more interesting. One thing I can tell you for sure, we never talked about television. A lot of these distanced friendships faltered after two or three exchanges, but many lasted for years. The key factor for me was whether or not the person could talk. And, of course, there was the value of information exchange. Did we share interests and could we mutually benefit from an exchange of information? Thinking back on to those early tapespondence days, I remember a few particularly valued tapespondents: Don from Riverside, California; Louis from Miami, Florida; and Jon from just outside London, England. Don was into heavy descriptive detail; Louis was into overcoming sexual frustration; and Jon was into tropical birds that talked, motorcycles, and shameless lying. Making video as an adult is a continuation of these early experiments in personal tapespondence.

I'm reflecting on my development before I realized that art had a place in culture. For me, culture did not include art until I left my small town for the city and eventually art school. This experience is, of course, still a reality for most Americans and Canadians. For the vast majority, culture is nothing more than the American movie, television, and music industries.

As an artist and writer, I have always been especially intrigued with authoring descriptive prose — with creating mental images as a concrete, though immaterial, manifestation of ideas. Each reader or listener can remake a text or spoken image from the parameters of a single line or phrase. It excited me that mental images were not confining the way physical, materialized images always were. Writing, the way language could be constructed in elaborate, extended sets of visual symbols, was obviously a visual art. When I set out to teach myself how to write, I first studied grammar and semantics, to learn the basics, with the goal of learning how to write a particularly visual descriptive prose. From there, I realized that written language was multisensory and that effective description of mental imaging was dependent on voice — the natural, fluid form of instructions as they rolled off the tongue. Following a lengthy study of writing by linguists, I came to the conclusion that I would have to learn to hear my inner voice if I were to write effectively. Strangely enough, although I found it easy to create vivid images in natural conversation, I couldn't hear or get the feel of my voice when I would sit down to write. I had somehow grown up thinking silently. I began to listen for my inner voice, but couldn't hear a whisper.

When trying to hear something very faint or possibly nonexistent deep inside, you become extremely sensitive to the noisy environments we all live in. I began to experiment with sensory deprivation, seeking an absolute, clinical silence. After experimentation with anechoic chambers and wilderness environments, I found a simple, practical way to cut off even the most subtle distractions of environmental noise. I bought a pair of skeet-shooting earmuffs. These protective gun-muffs resemble a set of stereo headphones, like the protective headgear aircraft-maintenance personnel wear when servicing jet aircraft. Late at night in the beautiful vacuum of my totally passive headset, I sat at my keyboard listening for my inner voice. I started "talking" to myself in my mind. I could hear my mind's voice, which was faint and generic compared to the acoustic character of my speaking voice. I found I could manufacture other people's voices with greater amplitude and truer fidelity in my mind. I assumed the difference between my weak generic mental voice and their rich acoustic voices lay in the difference between real-time inner-voice creation and playback from memory. I then found it relatively easy to synthesize specific, memorized voices with my real-time, self-initiated inner speech. For instance, I could speak forcefully in the voices of Bob Dylan, Jim Morrison, or Lou Reed. Associating with these prominent voice artists was problematic in that their characters (lifestyles, psychologies, politics) tended to influence the content of my internal monologues. I found that music, especially my favourite music, which had been forged into my memory through endless exposure at high volume, was easy to reproduce in my mind.

After months of regular internal-listening sessions (through device-based, reduced-stimuli meditation), I found the amplitude and clarity of my inner voice was increasing steadily. It was basically a question of discipline. As my attention span and ability to concentrate increased, I found I could speak to myself mentally in an "acoustic" voice. It wasn't exactly the same as hearing myself speak aloud, but it was precisely analogous. I concluded I could hear myself forming mental speech the way others would hear me speak. This inner voice was given better psychoacoustic definition when I began a meditation period by listening to my own recorded voice on a tape recorder.

Now that I was able to speak to myself, internally, I became increasingly dismayed at the banality of my internal monologues. While drastically limiting outside stimuli to hear my faintly transparent inner voice, I found I didn't have much to say. These exercises revealed my absolute dependence on external stimuli. One's speech feeds on the speech of others in conversations and one's inner voice normally converses with the multisensory barrage we receive every moment of every waking day. Just as we converse with others within a context of visual, acoustic, tactile, and olfactory information, we also speak to ourselves, internally, about this context. The problem is we are seldom consciously aware of our continuous inner monologues. I cite this lack of awareness as a problem because the inner voice and its position within the context of reality is fundamental to establishing and maintaining a unique psychological identity. In an age of electronic media, where television engineers conformity, it is useful to have a grasp of one's unique inner voice as an instrument of survival.

Even though I could hear my newfound inner voice in the silence of my skeet-shooting earmuffs, I lost this capability when I listened to music, even exclusively instrumental music like jazz, classical, or muzak. I craved acoustic input to respond to, to improvise within, but found myself completely distracted and overwhelmed by it. I tried watching TV with the sound off, experimenting with various types of programming, and found the most success with strict, redundant, fictional formats such as sitcoms and soap operas. Of course, there was a price to pay, as my inner voice inevitably settled into narrative structures conforming to the visual stories with their repeated sequences of formula images. When I returned to exercises in radical sensory deprivation, I discovered a way to stimulate internal images through the input of pure light. By cutting an ordinary ping-pong ball in half and placing the two halves over my eyes, I could bathe in a rich field of white light, while facing a powerful source such as window light, fluorescent, or incandescent lights. In the science of sensory deprivation, this phenomenon is called a *ganzfeld*: an even field of white light. With my ears covered by soundproof earmuffs and my light diffusing ping-pong ball goggles in place, there were freeflowing showers of internally generated images, which could be described and discussed by my clearly audible internal voice. Although these visions were often disturbing and violent, I had clearly found an effective personal methodology for generating my own descriptive voice prose, a way of composing visual art based on internal, highly personal visual experiences. After several weeks of productive imaging sessions, I put my fingers on my typewriter's keyboard and learned to transcribe my inner voice simultaneously into long, rambling texts I would later edit or excerpt for public exhibition as visual-art texts.

> > >

Coming up with the raw material of one's art in a vacuum is one thing; structuring the work for exhibition is another. The act of exhibition is an attempt to reach an audience, and when an artist is preparing to exhibit, he or she thinks a lot about audience. Making video art for an audience differs from making art in more traditional fine arts media mainly because of its exhibition context and potential audience. Video, like all fine arts media, can be shown in galleries and museums, but it can also be broadcast or cablecast. With television, the potential audience is remarkable in its scope and diversity, but it has also been conditioned by the conformity of available programming and the television viewing process itself. Video artists can't help but think about this audience the way broadcasters or cablecasters do. Unfortunately, this is a numbers game usually played by show business executives, not by creative people. The people who make broadcast media develop an abstract profile of the audience they serve. Broadcast artists, no matter how sensitive they are to an audience's needs and desires when they begin to produce a program, quickly fall into the trap of dictating public taste as self-appointed trend-setters. With television in particular, there is a tendency to want to control and manipulate the audience, to take the lead. It is possible to dominate the audience completely through a rich, fast-paced sequence of images in combination with conventional (beginning, middle, and end) storytelling, usually with a mix of dialogue and voice-over. Appropriate pace and emotional sweetening is accomplished through functional music.

Conventional television programs leave little room for the audience's inner voice. All television viewers have an inner voice, often running as an inaudible subtext, but it is little wonder it is so faint and underdeveloped. After all, television programming is like all software — it is a set of instructions to be performed by an audience — and a television audience is essentially a kind of machine, a biological machine with psychological and social characteristics. The problem with evaluating the performance of television software is that audience feedback is always cast in the past tense. You only get a feeling of how a program is being read by another at the moment it's being read, through facial expression and body language (though a television viewer's body is generally static). Such intuitive assessments need external reinforcement. Although more immediate and unobtrusive to the ongoing flow of programming than a verbal response, even eye contact reflects the immediate past.

Most viewers watch television in silence, without making external comments. If a program is engaging, even short verbal comments are often inappropriate, as a running commentary destroys the enjoyment or information value for others.

Wireless remote controls, now standard equipment, encourage channel hopping and have eliminated the frustration of being held captive to dull programs. The numerous advertising breaks, which once offered a chance for group commentary, have been effectively eliminated as remote channel-switching permits viewers to keep in touch with multiple programming sources, demanding verbally passive, continuously silent viewing. Discussion of TV content is thus maintained in the review mode, distanced by time. Notable terminology is passed around quickly, so that cartoon character Bart Simpson's "Don't have a cow, man!", for instance, becomes common usage for a generation of children.

Most individuals watch television alone, both literally and figuratively. If they have viewing companions, or recall their experiences later with others, they are still distanced profoundly through the strict convention of viewing silence. TV viewing is an intimate experience, differing significantly from the tradition of theatrical cinema. The movies offer a distinctly social viewing ritual, where large crowds meet to experience a big screen display in darkened movie theatres.

Home-video movie rentals have diminished the power of both cinema and television. Cinema as an art form has suffered as its screen and social dimensions have shrunk, while television has been hurt as a commercial information commodity, its once monolithic audience blocks now fragmented by the availability of more and more sources of cinematic programming. Individuals now watch both television and cinema in a greater silence, in further isolation, not only because both forms are consumed in the home and our experiences with the small screen do not generate much conversation, but because there is a greater diversity of programming for the small screen. Progressively, audiences have less and less identical media experience to share. In terms of programming diversity, television is devolving into radio, but the larger problem of viewer passivity and the underdevelopment of individual voice go unaddressed.

Television news, especially in its capacity for determining "live" experience on a global scale, has a predominant position from which all other prerecorded, reactive, and reflective media dangle, as appendages. Prerecorded programming can be thought of as events in storage, as future primary sources of psychological experience, but the influence of canned programming pales when compared with the power of "live" television events. Despite the undeniable zenith of "live" television, there is still a numbing sameness to most mediated experience today. This sameness is largely due our reliance on the rigidly uniform display of television screens and speakers, even for diverse information sources: perceptual engineering through a ubiquitous electronic instrument. The VCR has transformed the television receiver into the video monitor, now the new site of cinema through the phenomenon of video movies. Through time-shifting, we can even turn "live" television into prerecorded video information, to be replayed and shared with others. Artists, like everyone else, must negotiate today's reality, recognizing that our societies are constructed around television. As individual human beings, existing as genetically based, experientially shaped biological entities, we must pay attention to the nature and influence of television. Television is part of our natural world.

> > >

By 1977 I had come to the conclusion that television's audience was not as passive as those critical of television had presumed it to be. In fact, there was a second, inner voice providing all viewers with a personal running commentary, no matter how glazed over they appeared to be. This was a psychological activity that paralleled something else people were likely to be doing when indulging in less demanding media consumption, like radio listening. "Television's Human Nature" was an attempt to describe the internal monologue I knew was running in every viewer's mind, the inner voice of which they were generally completely unaware. My strategy for creating awareness of this internal voice was to meet the viewer head-on in the cablecast-broadcast context, focusing my narrative on the viewing process itself. "Television's Human Nature" was produced through a community access cable channel, Rogers Cable 10 in Toronto. It ran as a regular half-hour program for months within the programming spectrum of the twenty-plus channels Rogers offered its subscribers in 1977.

The format of "Television's Human Nature" is minimal: a static portrait of the television audience, where television viewers see a model viewer watching them from the other side of the screen. Against the conventions of the fast-paced dramatic narratives that dominate prime-time television, I offered a mirror image of the viewing audience's static viewing posture, including an audible internal monologue. A voice-over, the scripted internal monologue of the model viewer (a prerecorded voice performance by the on-screen viewer) was offered as the program's narrative.

By presenting a completely static television image, I hoped to accomplish two things. First, the radical stillness of "Television's Human Nature" was an arresting and unlikely image in the busy cable environment. This violation of television conventions focused attention on the scripted, simulated internal monologue. Fast-paced programming is designed to deflect the viewer's attention from his or her self rather than to make the viewer conscious of the situation. "Television's Human Nature," by contrast, turns viewers back on themselves. The voice-over is an internal muse on the television viewing process itself, comparing it to radio listening. In fact, the viewer of "Television's Human Nature" was indeed listening to radio, figuratively and in actuality, because the only moving images were mental images created by the internal voice of the model viewer. As viewers became self-consciously aware of their physical passivity — a kind of literal captivity — they were likely generating their own internal commentary alongside the "internal" monologue provided by the model viewer.

"Television's Human Nature" is passively aggressive television. A stranger sits silently watching the viewers watching television in the privacy of their living rooms.

As a television program, "Television's Human Nature" has a unique structure and complexities. It consists of three approximately seven-minute viewing periods by a male and female viewer, separated by a two ninety-second segments (cut in like commercials between viewer monologues) of "live," real-time television surveillance of a hostage-taking incident at a bank on Yonge Street. A man from Calgary had taken thirteen hostages in a Toronto Dominion bank across the street from the production studio of Rogers Cable 10. Rogers's staff simply stuck a camera out the window, framing the entrance to the bank and the police SWAT team crouching on either side of the door with shotguns, cablecasting the real-life drama "live," and scooping all the local broadcasters who were prohibited for security reasons from gaining access to the site. These highly charged, commercial-length, hostage-incident "snapshots" directly confront the viewers of my program with being a captive audience, amplifying the aggressive, confrontational stance of "Television's Human Nature." The first and second viewer portraits are straight, mirror-image confrontations with audience members. The third monologue is more entertaining, though also more demanding in certain ways. I use the device of the video dissolve to alternate back and forth between a male and female viewer,

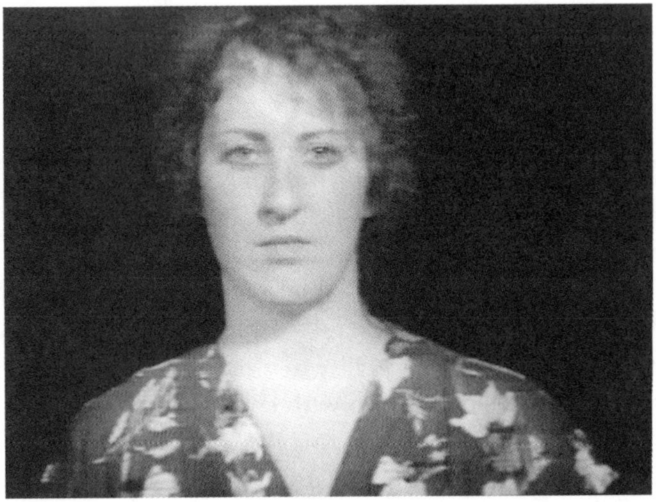

< < <

Susan Harrison as she appeared in Tom Sherman's video, "Television's Human Nature," 1977, 28 minutes and 30 seconds.

mixing their two voices alternately as foreground or background voices. In other words, the audience for this program finds itself listening to two simultaneous, identical monologues, differing only in gender, while adding its own internal response as the third voice in the video.

The programmers of Rogers Cable 10 were pleased with "Television's Human Nature" and aired it in all available time slots for months on end until their master recording literally fell apart. I was pleased in turn to have provided a mirror for their cable subscribers, and that my mirror was of a design that made television viewers self-conscious and aware of their viewing process and their internal monologues. In subsequent video works, I tried a number of different program formats, but I generally continued to experiment with visually static presentations, choosing to focus on oral narratives to force the audience back into itself and into an awareness of the latent inner voice locked within the viewing paralysis.

> > >

In the 1970s, a huge quantity of criticism was published about the disturbing passivity of the television viewer. Television's audience was depicted as a great mass of unthinking consumers of the values of late capitalism — an essentially voiceless society poisoning itself with a gross diet of commercial television, responding only by rushing out to buy products it was told it needed through criminally manipulative advertising. Most of this early television criticism developed from and distorted the formal insights of Marshall McLuhan, who coined phrases like the medium is the message, we are living in a global village, the word is dead, television is cool, radio is hot, and so on. McLuhan certainly had his doubts about the ultimate value of television. As a literary scholar and philosopher, he knew that television would diminish the power of the printed word and alter forever the definition of literacy, but the brilliance of his early writing was his ability to comprehend the effects of television, not to grind his readers down with conservative, nostalgic values. Post-McLuhan television criticism has largely defined the medium as the purveyor of social and psychological disease. People were addicted to television. They were alienated from society by television and reduced to illiteracy. The nuclear family was being systematically destroyed by TV. Gender stereotyping was refined and perpetuated by the medium. Our societies were becoming more violent, and our species had completely forgotten it was part of a biological ecosystem. Television was evil because it was controlled by rich people with only one goal: to make more money at the expense of everything else.

Any thinking person would acknowledge a measure of truth in these claims, but since McLuhan's death, little research and theory followed his probes into the nature and effects of media, nor have his theories found a broad public. In other words, how does the medium affect the audience in terms of visual and acoustic perception? Just how passive is the television audience? How is television affecting other media and their audiences? Unfortunately, the only people who have taken these questions seriously and responded intelligently are advertising people, psychologists, and media artists.

Advertising people do some serious thinking about these questions, functioning almost like scientists and then applying their findings in updated, state-of-the-art manipulations of television's audience. Perceptual or cognitive psychologists conduct their research within the confines of academia,

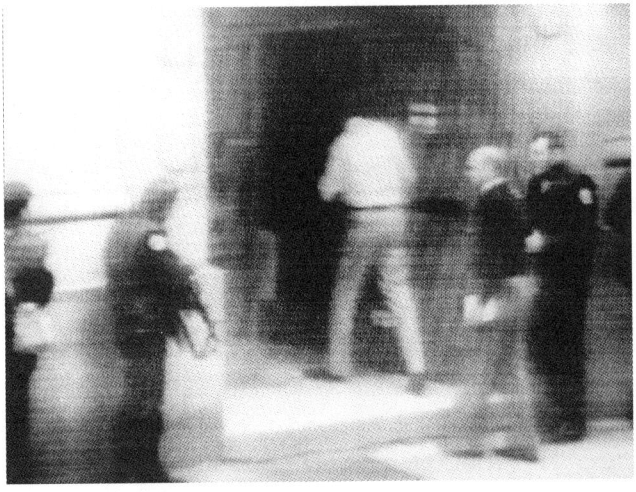

< < <

Police negotiators enter Toronto Dominion bank on Yonge Street in Toronto during hostage-taking; a scene from Tom Sherman's video, "Television's Human Nature," 1977, 28 minutes and 30 seconds.

often under government or corporate contract, governed by nondisclosure agreements. Their research findings, when not reviewed and applied in secrecy, are published in obscure academic journals. Media artists make media-based art, often produced directly in video, which is inherently critical of television and can be described as antitelevision. Unfortunately, this antitelevision has found limited audiences to date.

Radio, of course, is still with us — and it is arguably the only serious media rival to the powerful leading role of "live" television. Radio continues to emanate simultaneously from the site of events of global importance, holding a considerable edge in the coverage of regional and local events. Radio is closer to the street, and, in its integration with the telephone, is composed literally of the voices of the people. It is instantly updatable and does not suffer from the same scheduling restrictions as television. Radio is not in competition with cinema and does not demand the full attention of its listeners. It functions independently of moving pictures and can be consumed while its audience is doing other things. The prime times of urban radio are determined by traffic flow, when the largest number of people are in their cars, driving to and from their homes and their television screens. Its prerecorded content is mostly music of relatively short duration and all its canned programming is continuously interwoven with "live" commentary. The diversity of television and video, only recently increasing as a result of cable, satellites, and the VCR, has a long way to go to catch up with the plethora of information available on the dial of a multiband radio receiver. Radio is a rich blend of information sources, unbound to concrete images. It is composed of many voices, the speech and singing of voice artists, and the voices of others who choose to speak out in public. Radio is the most comprehensive media representation of our oral culture, the mediated locale of people aware of the way they sound and making sense when they speak.

Radio is thus a state of mind distinct from those of television, cinema, and video. It is a companion in a conversation with reality, a full interaction with the world. We may sit and listen intently to radio, ceasing all other activities if its programming deserves our complete attention, just as we must do with moving picture media, whether or not it is worth our while. But most listen to radio while actively engaged elsewhere, using it to enrich their experience with the real, unmediated world. Although it is technically true that television is radio with pictures, as it was in its earliest days — listening to television today without paying attention to its pictures is a rather poor "radio" experience, an abstract, generic companion. Radio itself, though stiff from time to time, is far richer in voice and more fluid in terms of thinking.

News summaries on both radio and television are written in advance and differ little when compared as voice. But where so much television is scripted as theatre and cinema (with the exception of talk shows, interviews, sports, or game shows), most speech on radio is improvised as monologue or conversation. When watching fictional television (or even TV news), one reads a sequence of images while listening to the voice performance of a script. Visual decoding of all motion picture media is similarly an act of reading, which demands a relative state of physical passivity, and hence a static viewer. But listening, while requiring focus in its most demanding moments, offers freedom enough for real mobility, even dancing.

Apart from its programming diversity and high level of information, my interest in radio is based in its direct relationship with the spoken word. I have intentionally used video (and television) as sites for radiolike monologues, transmitting voice narratives into the consciousness of static, passive viewers. To this end, I have slowed the visual narratives of my video works to a standstill, violating the conventions of television programming. With the reduction of complexity and velocity in visual structure, I release the viewer to listen. Often, as with "Television's Human Nature," the viewer's silent "reading" process is confronted, demanding a self-consciousness bound to make the viewer uncomfortable. Such works have been criticized as "boring," a charge to be expected when the conventions of entertainment and engagement are violated. My strategy has been to force the viewer back into his or her self, rather than taking the viewer out of his or her body into another dimension, making time fly.

Digital technologies and the new potential for interactivity came to prominence in the 1980s, with effects on input for both information and narrative. Artificial intelligence and virtual realities were on the horizon. I began to fantasize about a machine companion that could watch me and understand my natural language, my speech. In 1986, I began a body of work synthesizing my interests in visual iconic languages, pictograms, and ideograms, and image-and-voice recognition systems. I structured my theories as visual icons and began a script — a set of programming instructions — to bring this machine to life.

Eventually, I integrated all media technologies into one hybrid machine entity, or "smart" robot. In those days there seemed to be a general anxiety in the workforce about replacement by "smart" machines. Pushing that idea further, I proposed an intelligent machine-companion, a preoccupation for years to follow. I would create the robot of my dreams, or more gently, a machine with typically human attributes. It would get tired and have to rest, and it would make mistakes. It would sleep and dream, contract diseases, and slip into chemical dependencies. It would have limitations, like any human. Parenting such a creation intrigued me.

"Exclusive Memory" (1987) was my first foray into developing an intelligent computer-based robot entity which saw the world through video inputs and heard everything within range of its microphones, via my video recordings. The work was produced while I was artist-in-residence at the Western Front, a video-production facility in Vancouver. My original plan was to sit in front of this machine for a period of eight hours (to coincide with a natural human period of uninterrupted sleep), improvising a conversational monologue directly to a pair of cameras (two eyes, two points of view), and two microphones (two ears), the sensory input components of my machine companion. I wanted to create an immense sea of words organized and anchored by a series of static visual scenes or listening contexts. In rehearsal, however, I found I could only maintain a coherent monologue for thirty to thirty-five minutes at a stretch. With these new parameters, I recorded in twenty-minute segments, talking to a pair of video cameras (and microphones) until each of eighteen cassettes was fully recorded, a total of six hours. Each section has its own look and feel. The two cameras and stereo audio were mixed live during the recording process.

These plans had gestated for over a year, and when I sat down to make the "experience transfer," the words just poured out. Although I play myself in this performance, telling the machine that I am Tom Sherman, my character is fictionalized to a degree. I said I was a university scientist, a researcher in artificial intelligence, and the creator of hardware and software for a human-like smart robot. It was my offspring. In my visual appearance and spoken monologue to this machine, I undertook to parent it, giving it knowledge about the world from my personal perspective. I spoke about the things we had done together, reflecting on experiences with our family in Nova Scotia. I recounted how I had flown back to Vancouver with my machine, ahead of the family, to return it to the university lab, where technicians would upgrade its memory and image-processing software. From reflection on our common activities, I expanded with other stories to enrich its base of experience. While its "body" was at the university, I was directly engaged with it in real time (as it were) via remote video and audio feeds from my Vancouver home. In this state of electronic intimacy, I hoped to become one with my machine.

Even though this construct was perverse, I found I could perform within the person-machine relationship; I was comfortable with the situation. After all, I had been preparing for this relationship my whole life. For years, I had felt under surveillance by the performers and producers of the television I had watched; I had listened and responded in my internal voice. I had maintained friendships across great distance, over time, through audio and video tapespondence, and countless telephone conversations. I had been in rewarding person-machine relationships for years.

The voice I used to talk to my robot was an externalization of my inner voice. I was talking directly to a machine, not to an audience through my machine. My machine would compile my words phrase by phrase, inheriting knowledge about my character as my monologue progressed. As it learned more and more about me, hour by hour, I found my voice being altered by the process. My intention was to create another individual intelligence, not a clone of myself. The robot was demanding in its passivity.

My machine had become primarily a listening device. Formally, a moving picture medium was reduced to a static image form. I rationalized that my robot had a limited comprehension of visual motion, that its voice recognition and comprehension circuits were far more advanced than its visual apparatus. I sat motionless in front of my machine, limiting my motion to my lips, facial gestures, and hand movements for emphasis of my speech. I was aware that my machine's sensory limitations were a particular condition of sensory deprivation. When I reflected on our shared experience in Nova Scotia, I gave the machine mobility, a kind of advanced, self-generated portability, but concentrated on its memory of static objects, like dead animals. This consistent reflection on motionless sensory experience (in memory) combined with the "live" immobility of its parent, Tom Sherman, provided a more or less surreal impression.

The dream state, while not necessary static, is the result of our shutting down most of our sensory input during sleep. We digest and restructure our prior experience through the process of dreaming. My machine entity was

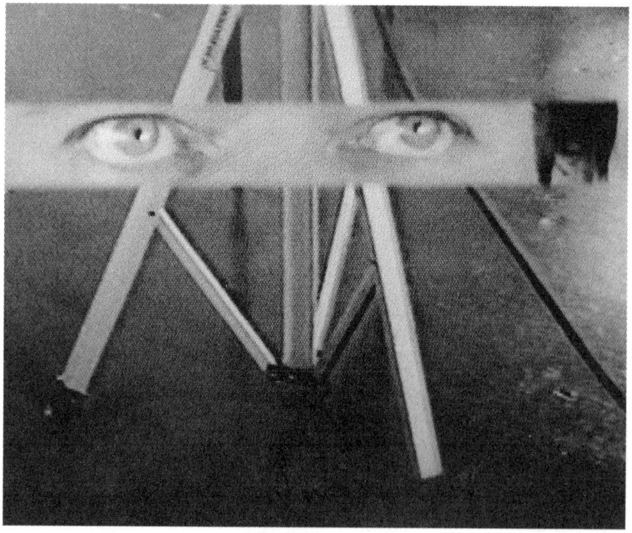

> > >

Tom Sherman, relating to and becoming a machine entity in his video, "Exclusive Memory," 1987, 182 minutes.

a "sleeping robot," recompiling its memory as it consumed my voice narrative, a fresh "live" experience, a performance largely devoid of visual motion except in the flow of my continuous speech. My natural language fleshed out my static image.

During my hours in front of the cameras, I believed I was talking to a machine, my robot. I hardly felt the presence of the other, secondary audience, the people who would eventually watch and listen to my experience transfer. It was never my intention to treat my eventual viewing audience like a machine, though of course this slight happens in the end. My objective was to speak the way a lone viewer talks to television, in the form or his or her internal monologue. I externalized this inner voice, using the privacy of my relationship with my machine to advantage. I was actively speaking, out loud, with my lips moving. It was my own voice. I was speaking my mind, off the top of my head. Sure, I was posturing, self-conscious of the image I was projecting. I wanted to make a good impression, for this performance was for my direct offspring. Yes, I sometimes censored myself and stretched the truth. When I decided to make part of my disclosure public, by editing the finished monologue for public exposure to the other audience — the people who would eventually observe me in my relationship with my machine — I left exactly half of my performance on the shelf. The three hours of recordings I have never shown to the public remain in the exclusive memory of my machine. I felt some comfort in the privacy afforded by the scope of my public disclosure, as the final three-hour version of "Exclusive Memory" is so immense as a narrative structure, it is difficult to get a handle on.

"Exclusive Memory," in its final form as a three-hour video monologue, is essentially a radiolike experience. Without its video information, my disembodied voice would have a diminished relationship with its machine. Yet "Exclusive Memory," like "Television's Human Nature" before it, highlights the mental voice, filling the void between psychological reality and the unreal nature of television. Television, even transporting us to the time and place of "live" events, is less impressive when we are unaware of the commentary of our internal voice. Radio offers a point of view less restricted than the voice welded to a television image. Radio permits the expression of personal and cultural identity, floating through the air in the freedom of visual discontinuity. The video images of "Television's Human Nature" and "Exclusive Memory," static and unchanging, become fixed sites from which my voices emanate.

With "Television's Human Nature" and "Exclusive Memory" I turn television into radio, employing the visual iconography of stationary viewers fixed within the video frame, whether or not they move their lips or talk out loud when they address their audience. These video recordings are designed as software for generating and amplifying self-authored, self-conscious monologues in the minds of all those who experience them. My objective is not to distract audiences through the conventions of storytelling or to mesmerize them through visual spectacle. I simply want my audiences to turn up the voices in their minds.

The Artist Attains Ham Radio Status in an Era of Total Thought Conveyance 1978

000001 The artist attains ham radio status in an era of total thought conveyance.

000010 Or the artist becomes a radio-frequency engineer, fully licensed to build the bridges for their minds to cross.

000011 We now turn our attention to the pioneers of the personal communications field.

000100 By a twist of fate, they were among the first people to get their hands on the equipment.

000101 They were able to log their private lives as worthy material for public display because of their early entrance.

000110 Later on, those working in the new medium would have to approach the art of revealing themselves in an entirely different manner.

000111 If they remained natural in their relationship with the technology, they would be labelled derivative by the critical community.

001000 For a while there, everything they recorded was interesting, not only to them, but to others as well.

001001 All they had to do was look and feel different than the television they were already receiving from the standard broadcast world.

001010 After a decade of being outside the look and feel of regular broadcast television, they would make their move of contradiction.

001011 They would make the new broadcast television.

001100 For they had learned television working under their own set of rules, building confidence through years of small production schedules, until they reached an age where they believed their television to be worthy of its final destination, the standard broadcast channels.

001101 They declared themselves competitive entering the commercial broadcast field.

001110 They found their aim to please.

001111 They made their statements to the press.

010000 They, too, were in it for the money.

010001 If the artist finds the opportunity and accepts the responsibility of programming broadcast television, must this responsibility be an enslavement, rendering the artist powerless, lost in total servitude to the general viewing public?

010010 Must the obvious potential for directing the thought of the mass audience be lost in the shuffle of this most difficult task of producing art for mass consumption through television?

010011 Are we merely presenting the artist as a contestant in a popularity contest?

010100 Yes, by all means, artistic control of the work will be hard to maintain.

010101 So, as the new material is presented to the audience, as this thought is externalized over and over, undergoing a certain natural clarification through unavoidable redundancy, it should become clear in time, that the television artist has different concerns, high-profile, attractive concerns, whether or not these concerns presently appear extravagant or too visible to be true, or whether it shall turn out that these matters are but a prompter's words encouraging much more factitious actions than this or that one. Writing this — broadcasting that.

010110 But all this and that is only so far, and therefore it cannot be nearly all that is true.

010111 It is up to you to make sure.

011000 Am I not jumping ahead too far?

011001 For if artists actually were in the thick of it, they would be looking and feeling different; yet another full turn around, they would be talking different, producing more, getting tighter all the time.

011010 For instance, their behaviour on the set, on the stage, time after time, the directors would emerge from their conversation with the working technicians where only equipment was discussed. The down-to-earth language of the technical class. The jargon in the lounge, the silence in the control room.

011011 Without exception, the die was cast, time after time, the producers would never question themselves as artists. All they would want was the best possible results from the crew, communicating through the director, making the decisions that would structure the whole, or more simply put, it was their job to push the program through the budget.

011100 Freelancing for the institutions that have nothing to say, won't you find it restricting? ... their collective mental state shrinking skull-tight around your free thought, thereby canning your imaginative nature.

011101 How great is the need for personal message on the major networks?

011110 If communication is control and the degree of conveyance provides the basis for power, chances are they're not going to put the individual on mass television simply because he or she thinks well.

011111 The controlling members of the communications industry have good reason to believe the artist will use the first opportunity to be seen as well as heard by a wide audience while performing a sensational masturbatory activity.

100000 Colourful behaviour increasing the amplitude of the cliché.

100001 The artist making television as the artist is expected to ... thrilling shots bleed white hot burns across the surface of the tube ... making television as the artist is expected to.

100010 The formal gesture of the single individual divided by millions of monitored shares.

100011 The staggering division of a man or a woman crying aloud with the pain of multiplication, the fact of the matter of being duplicated more than a million times by every kind of receiver after having said only they, the people, want me.

100100 They, the people, want me to cry for an hour.

100101 So quickly broadcast television turns into just another performance situation for the artist.

100110 The picture of the family huddled down low posing in front of the transmission equipment; this picture will last but a short time.

100111 Having something to tell everyone on television. Tell everyone something by calling on the open channel. Showing your face. Exposing your thoughts. Talking to no one in particular. Holding on to the air. You can listen in. Watch me. I've got a message for you.

101000 For a television platform, we've got ideas. A platform, a structure, if only a collection of thoughts to be executed within solid visual order; have we got ideas? You can have your little spit-bubble of a world. But I'd like to tell you what I see in the tear-drop windows of your moving walls, your screens open to my cameras, your lips, your eyes tell me everything.

101001 Am I wrong to think I can get the message across with sex? Just a kiss, shit. Fuck the sponsor; roll them back and put them on pause; they'll have to hold their breath a minute, I've something to say. This is a high-priority public media address. A crucial, very loud announcement. An AV blast of the absolutely necessary truth.

101010 Television time, all air time, it's measured by their attention span, not by the hour or half. All of it, the air, it's measured by their thoughts. It's an air loaded with their sympathies. Or is it limited by their sympathies? You control the air when you're on the microphone or the camera, limited as you are as an individual. As an individual, you must box with their strict control. You should charge them for lack of information, for personal losses suffered under the present low level of thought conveyance, for your embarrassingly light conversation under their strict control.

> > >

The binary code numbering each text fragment, one through forty-two, was used to indicate the increasing role of computers in video-television culture. These text fragments can be shuffled, by hand or using a random-number generator, thus forming interchangeable, varying developmental narratives. In other words, artists will interact with technological media in similar but different ways, depending on a number of behavioural factors, including the evolution and advance of digital technologies.

Appearance, Memory, and Influence 1986

Appearance is important. The way you feel on camera makes a difference in the way you look. The clothes you wear, the way you wear your hair, the way you treat the camera and the person behind it. Your eyes tell us everything through a close-up. Your walk forces us back as your boots step firmly into a circle of brilliant white light. Flat black boots in silhouette against the glare of the studio floor. One of the toes turns up into the profile of a snarling rat. The other turns away, lying flat, hidden in shadow. Silent, lurking, potentially diseased. The monitor in the control room provides us with an immediate perspective on an imaginary video environment. The director calls the shots in this virtual reality. In video, appearance is important. In this first scene, a slick pair of boots turn out to be a couple of rats. One is aggressive, menacing. The other is playing possum.

You might wonder, what in the hell is the author up to? The artist who strikes out with such an unlikely image at the outset of an essay cannot be very serious about communication. Fuck communication. Nobody needs communicators. Communicators don't exist, except in the minds of those who wear starchy, buttoned-down collars and more than the occasional turtleneck. If the left/right-brain hypothesis makes you want to puke, then you too know the academics are anything but communicators. You understand their suppositions but reject them because you simply can't relate. They say the left lobe is linear and rational and superstraight, while the right side of the brain is wild and crazy and prone to being naked and glistening and as slippery as liver and deep red like a beet; not grey, but a soft, blood-wet organ framed as an illusion in a cheap mirror with a hand-sewn leather frame. Honest to God, the redundancy of their high-pitched, whining chorus of ignorance induces one massive, globally comprehensive tension headache on the intellectual populace, thinly scattered and remote as it is. The neuro-cultural chant of the McLuhanite drones will likely continue to build relentlessly until every electronic mirror on the planet implodes in sync with death-defying intensity and penultimate misfortune and disgusting waste.

What's the point of forcing honest people to analyze a culture of apparent lies? Whether they listen to those who speak openly in an unguarded, natural tone of voice, or to those who deceive through elegant, subtle, devious manipulations, the problem for the audience is always, ultimately, boredom. Normal tendencies in dress and fashion, or in the generic choreography of behaviour, are naturally redundant and inevitably depressing in their predictability. The stereotypical personality is certainly a fundamental aspect of human limitation. Media coverage merely exacerbates our insecurities, inviting us to stretch the truth. In a culture of deceit, the surface dips and swells with changes of appearance, both radical and gradual, all of a sudden, day by day, once in a blue moon. A belly laugh is toned down, a cigarette fetish eliminated, the hair flips up and disappears into the stiff brutality of razor stubble. The cameras play their part, and the viewers have little choice but to sense the apparent differences and to respond in kind. Video appears to accelerate the frequency of change. Or does it contribute to a tenuous stability by grabbing hold of an image or sound, gradually freezing our relative powers of association? The anchorperson reads the news every night inside the rigid frame of a single shot. Her goodbye smile concludes our daily thirty-minute picture of the world. Could we manage to wrap up the world in half an hour without her professional assistance?

The reality of television news is a manifestation of video. The newscasters come and go, while the world remains the same. Nobody cares enough to tape the news, though most will swear they want to save the world. The serious gap between stated convictions and actual performance represents the contradictory, gutless behaviour we have engineered through our schools. The fashionable posture of concern is tough to stomach in those who think there is nothing to lose. Maybe this empty liberalism is the reason our governments are at war with education. Whatever we think about the value of formal education, the children are the victims, because they can't find work and pay taxes, and none of the adults, not even the most perfect parents, want to foot the bill for their children's learning experience. Most people think it'll get even worse toward the end of the century. But let's flip the future around on its head for the purpose of argument.

In a much more optimistic scenario for the twenty-first century, let's imagine our children will all have jobs, and they will somewhat reluctantly pick up the tab for massive programs of adult education. Those over sixteen years of age, if they are smart, will volunteer for intensive training in video communications. Talk about potential growth industries! In fifteen years communication will be defined as stimulation through simulation. Fifteen years might be too far ahead to see anything clearly, so how about ten? If we are going to be prepared for the near future, then it's back to the basics for everyone. Fuck the ten-year plan; artists are already at least a decade ahead. Through the behaviour of artists, we may witness the turn of the century today. The artists, the only people living in the present, can't help but telegraph the future to all those who reside in the past. As obvious as this point is, the last thing the human race wants to be told is that it's dependent on artists to periodically introduce a few cracks in the repressive wall of obsolete thought and behaviour.

Real changes occur in individuals first, and artists working with a powerful technology like video are literally in possession of a superweapon of perceptual transformation. Video is the supreme technology of the era for sharpening or numbing perception. Video is the ultimate instrument of perceptual transformation, but quite frankly, unless artists know exactly what they are doing with this technology, their circuits are likely to end up overloading. In many cases that's the fun part, commonly referred to as the buzz. Humming a sixty-cycle tune of image intoxication, a real potential for unnatural sensitivity exists with video. It's too bad so few artists succeed in unleashing the full power of this instrument of perceptual transformation. Communicating perceptual change will never happen if artists insist on sticking all their perceptual tools up their own asses. But then again, communication may be best understood to be a specialized function of the communicators.

Video is a ubiquitous communications technology used to record and represent images and sounds electronically. It is a push-button technology, easy to use. Technically speaking, if the light and acoustics of a situation are well understood, all anyone has to do to achieve the maximum level of performance from a video recorder is point the camera in the right direction and position the microphone in the proper place. Video technology is reliable and portable and quiet and simple as A, B, C. It provides satisfying results instantly. Unfortunately, video technology is all too frequently used to record and represent the most pedestrian level of thought by people who see only what they want to see and never listen. Only rarely does the artist transcend the electronic doldrums of video banality. On the surface of the screen, the medium does appear to have a magnificent, largely unrealized potential. This unrealized potential is immediately apparent in the initial stages of exposure and disclosure. When an individual perceives his or her image through video, his or her appearance is a revelation.

With video, one's image is so different, so very far away from the flesh and blood you imagined. Watching yourself on the screen is almost like seeing oneself through the eyes of a stranger. Across the courtyard, you sit in your kitchen drinking coffee in silence. The soft natural light of the day plays across your brow. Your hair falls as you lower your eyes. You study your image as a stranger would. The frame of the window outlines the subject under observation. The time is repeated, the space confined. The movement is predictable. The image is stable, though the life form appears fragile from a distance. The body is so vulnerable in front of the mind. The hands are exact, the fingers articulate. The face changes like a signature, whether rushed or forceful or loving or curt. The human form resides perfectly within its limitations. The life form on the screen, this viewpoint from the mind's eye, is the gift we receive from the machine called video. By reducing the magnificent complexity of life to only that surface reality represented by video, the mere exposure of one's physical body to the camera becomes analogous to a full disclosure of heart and soul. Standing alone naked in front of oneself may be perceived as an act of love. The video machine grants us this privilege, such a unique relationship with ourselves. How will we ever succeed in expressing our gratitude to this machine?

Many may wish to acknowledge the value of these machines, but millions more might question the necessity of expressing our gratitude to the inanimate infrastructure of our civilization. Video, after all, is a technology engineered to suit the species' purposes in general and to some degree to respond to our individual needs. For most individuals with a typically superficial relationship with video, there is no need to give anything back to the machines, other than the intensity of undivided attention. We all take for granted the fact that a global population may witness an event or environment instantly and simultaneously through the common devices of the camera and microphone. All of us who love to watch appreciate the technology that has permitted us to look in on events and observe environments anywhere within the range of the technology's artificial point of view. We have only begun to examine with our senses a universe preconceived exclusively in abstract terms. Individuals bound by the normal prerequisites of human mortality can now appear to be in two places at the same time.

Although photography and film had excelled historically as the ultimate technologies of memory, video quickly rewrote our perception of memory through an apparent multiplication of space. A "live" event transmitted from one location to another through the devices of video can be described as a transient display of instant memory. Magnetic tape, like photochemical media before it, functions concretely as hard copy. Images and sound may be transmitted from machine to machine with video, with or without tape as a backup. In other words, the hard copy is optional. Video memory is fundamentally transient. We focus our attention on the tape as a physical medium, but video travels well beyond our senses at the speed of light, from machine to machine. Video transmitted as radio is called television. Television is frequently the transient memory of film prerecorded on videotape. Tomorrow's cinema will be just as fleeting. Cinema will be video without tape in a theatre, and events on the big screen will appear to be "live." Space will apparently be multiplied, and our perception of memory will be rewritten again by the technology of video.

Memory is a place in the distance where we observe ourselves. In memory, we leave our bodies for the space and time of our minds. Memory is a silent shower of light. The body suspended, the mind fastened tight. Memory is hard to control. We view the past as we must, in the best possible light. We second-guess the pain. Our minds are full of conclusions. Memory is recollection, and recollection is the trigger of imagination. The past is the truth in the shadow of doubt. The technologies of memory are often utilized as instruments of deceit. Memory is written and rewritten over and over again. Documentation is the sticky residue that binds together the loose ends of our imperfect memories. To verify all events within the broad context of reality, we must check and recheck the record. Our relationships with machines are documented by the recordings left behind. The shelf life of a video recording depends to a greater extent on our need to remember than on the environmental conditions governing the security of storage. The archivist must save everything if he or she wishes to be remembered at all. The good archivist will be remembered as a creature bearing both the wings of an angel and the horns and tail of a devil. The video machine has become our personal, inanimate archivist. Our intimate relationships with these machines are well documented.

Memory is fascinating, especially when it is framed by the artist-machine relationship. The dance the artist improvises in front of the recording machine is elaborate, deviant, and diversionary. The artist treats the machine with respect as the primary audience because his or her image is at stake. The precision and clarity of representation offered by the machine forces the artist to measure up. In each attempt to avoid displaying behaviour that might be considered mundane and stereotypical, the artist attempts to manifest a presence both challenging and acceptable to his or her audience. Take note that the machine is loyal, regardless of the artist's creative abilities. The audience of an artistic performance, whether the behaviour is displayed in front of or from behind the camera, is always presumed by the critic to be human, or at least an animate intelligence. The technology itself, which comprises the sophisticated, inanimate infrastructure of our civilization, is never considered to be the primary receptor for the artist's image. This critical, anthropomorphic bias obscures the process of making art while under the influence of the machine. Video is an instrument in the tradition of the telescope or microscope, except its power lies not in the science of optics, but in the realm of perceptual transformation. We do not see the world in greater detail, but we do sense the world differently with video. Space is multiplied on-line, while events are marked precisely in machine-time, microsecond by microsecond. The image of reality perceived while in sensory liaison with the video machine causes one to become extremely preoccupied with the nature of time.

Exposure to a technology as attractive and resourceful as video inevitably leads to intoxication and addiction. Withdrawal is painful and debilitating. Avoidance results in anxiety, followed by prolonged periods of depression. The influence of this dominant communications technology is exhaustive to the degree that many become hopelessly lost without the machine. Video dominates with its magnetism and authority. As a species, we must offer video something in return for our freedom. The servile artist must be generous in response to the demands of his or her technological master. The technology of video was conceived and realized by engineers for the purpose of extending our physical presence through a global, universal network of machines. Video technology has been perfected and employed to the degree that our sense of space and time has been radically transformed, conceivably forever. Turning our backs or burying our heads in the sand are not realistic options. We must face up to the problems we are having in our relationship with video technology. Most of us sense something is deeply wrong with the way we visualize the machines. We visualize the machines with only our images on the screens. We are constantly dislocated by our insistence on a video representation of the human form. Even when a landscape with another species is displaced through video, we seldom resist a deranged voice-over. Influence is dominance and mastery and control. We must find a way of expressing our gratitude to this machine called video.

With the marriage of video and computers, more and more artists are turning their attention to the image of the machine itself. The digital computer involves many artists in the origination of new images, both visual and acoustic, through the discipline of programming abstract computer languages. Ironically, the long-term goal of this form of image invention may be representation through simulation. In other words, writing instructions that permit the computer to draw representational imagery. On the other hand, a number of artists are focusing their attention on perceptual simulation with the machines. This artificial simulation of perception in both analog and digital machines is called image processing. Image processing is the manipulation of available images and the invention of new images through amplification, distortion, transfiguration, and synthesis. For instance, images from an analog video camera may be fed into a digital computer processor, which will impose its own order on the incoming video with the output or display becoming a thing in itself. Artists involved with such hybrid video-computer instruments are doing their part to create a different picture of the artist-machine relationship. The machine is finally permitted to articulate and display its own image. The artist has found a way of expressing his or her gratitude to the machine. The artist has offered the machine creative autonomy. We seem to be off the hook for the time being. Finally, the machines may influence each other freely, uninhibited by our selfish reservations and anthropocentric criticism. Only time will tell whether the human audience has the capacity to love the machine for being itself.

> > >

There are four primary acts of perception in artist/machine relationships:

1. The artist's perception of the world through the machine's sensors
2. The artist's perception of himself or herself using the machine as mirror
3. The artist's perception of the machine using himself or herself as mirror
4. The machine's image of itself, the artist attends and observes

The Art-Style Computer-Processing System

1974

As it is, our perception of things is a circuit unable to admit a great variety of new sensations all at once. Human perception is best suited to slow modifications of routine behaviour.
— GEORGE KUBLER[1]

A communication system is for sending and receiving messages. A communication system consists of two transceivers: two components that transmit and receive messages. A communication system is limited when the message is sent in only one direction, transmitter to receiver. This limited system can be expanded by integrating a processing system at the receiver end to provide access to the message. A processing system permits special treatment of the message at the receiver end of the communications system. The computer-processing system described in this text is specifically designed to manipulate the message transmitted to the two-dimensional surface of the video screen. The message source is a commercial television broadcast. The message is limited to display on the *flat* surface of the video screen. An analogy is formed between processing the video message and the act of painting. This processing system provides personal choice of how the message source is viewed, in the same way the painter chooses to view the environment through his or her method or style of painting. This system is labelled the ASCPS.

Style is a phenomenon of perception governed by the coincidence of certain physical conditions.

The ASCPS is constructed of information obtained from every major > 173
historically innovative treatment of the two-dimensional surface. The system contains the concise history of painting. By block encoding historically successful modes of sensing, the system contains a set of period visions. These period visions are methods of seeing the environment. They are rule-governed styles for processing messages. The rules are those instituted by

schools of painting dominating particular periods of history. At this time, period visions contained by the system are: Abstract Expressionism, Abstract Impressionism, Action Painting, Arabesque, Art Nouveau, Automatism, Barbizon School, Baroque, Bio-Morphic, Cartoon, Classic, Colour-Field, Cubism, Dada, Danube School, Divisionism, Expressionism, Fauvism, Futurism, Gothic (Late and International), Group of Seven, History Painting, Hudson River School, Impressionism, London Group, Mannerism, Neo-Classic, Neo-Impressionism, Optical, Orphic Cubism, Painterly Abstraction, Photo-Realism, Pointillism, Post-Impressionism, Primitive, Rayonism, Realism, Renaissance, Rococo, Romanist, Romantic, Social Realism, Super-Realism, Suprematism, Surrealism, Synthetism, Tenebrism, and Vorticism.

The vision circuit for each period contains additional processing characteristics. These simulate decisions of individual artists contained by schools of painting or period visions. The capacity of each period vision-processing circuit depends on sensitive patterning of physical conditions marking the consistent vision. Fine adjustment control of contrast, brightness, colour, and form open the end of each processing channel. The viewer fine-tunes a wide band of processed message, attaining authorship of the message. The commercial cable television system provides structure for immediate integration of the ASCPS in the home viewing system. The ASCPS consists of a centrally located computer with remote control units functioning as switching devices which afford access to the processing circuits of the system. Passing the message through a chosen period vision is accomplished by switching in the desired circuit by push-button selection on the remote control unit.

Application of the ASCPS: The message is the broadcast of a network news program. The information is in colour with low interference. The viewer decides to process the message with the period vision labelled Rayonism. Rayonism was an abstract Russian movement stylistically between Futurism and Abstract Expressionism. Mikhail Larionov, an instructor at the University of Moscow, published the Rayonist manifesto in 1913. This selection is made on the remote-control unit. While passing through the processing circuit, the message form is disintegrated to simulate the radiation of lines of force emanating from the objects in the news program. Important artists having this period vision are Mikhail Larionov and Natalia Gontcharova. The viewer chooses to understand Rayonism through Larionov's vision. This processing circuit is switched in through a selection made on the control unit. Larionov's vision is nonobjective. Visually, the news program is processed into a pure abstraction, with objects becoming new forms as they disintegrate into radiating colours. Fine-tuning controls permit control of colour and contrast with Larionov's radiations.

The ASCPS is introduced to provide the best possible system for the study of the history of painting. The system provides a previously unattained view of the artist's systematic attempts to attain efficient communication through the two-dimensional channel. The system simulates the collective visual experience of recorded history and offers choice of vision to the one-way communication system of commercial broadcast television. The viewer, in implementing historically incoherent methods of sensing on a contemporary message, can introduce message equivocation; that is, uncertain knowledge about the transmitted message when the received message is known. Uncertainty of message increases with degrees of image latency, the time interval between image and response or understanding. Message latency is abstraction.

The ASCPS is a stochastic system. That is, its output is in part dependent on random or unpredictable events. Total randomness of message produces monotony, a sense of sameness. Period vision-processing circuits pattern and structure the message. The structure provides a familiar visual language allowing new sensations to be perceived through contrast. By processing an available random message, broadcast television, the two-dimensional output of this communication system becomes a highly structured moving image with a degree of unpredictability.

The completion and integration of the ASCPS into the existing cable television system effectively surrounds (contains) the history of painting. Expansion of the methods of communication depends on technological invention. Components of this video-processing computer system are being designed and tested by technological artists in scattered communities around the globe.

1 Epigraph from George Kubler, *The Shape of Time: Remarks on the History of Things,* Yale University Press (New Haven and London), 123–4.

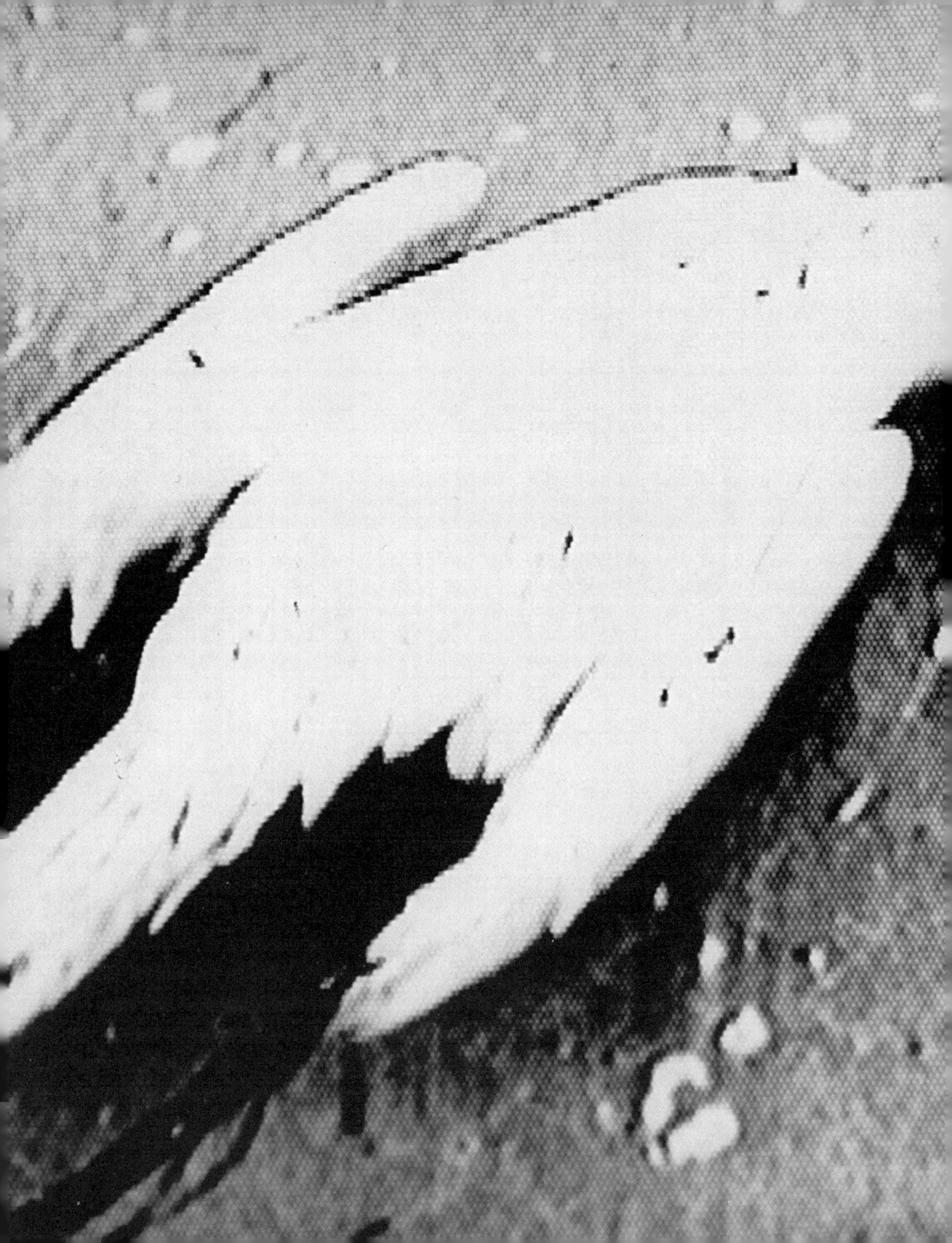

PROSTHETICS AND MEMETICS

Machines R Us 1994

[1] Machines R Us.

We love our machines. They are our friends. We find they are more trustworthy than people. But as we connect our machines with other machines, we find we are connected with people we don't know. We find we are being watched, listened to, observed, monitored by strangers through our machines. We don't like to be observed unless we call the shots. We want to direct every scene we appear in. We want there to be a clear distinction between when we are on the set and when we are off limits. When we are off limits, we are in our zone of privacy. We have a problem. We know that when we use our machines, when our machines are connected to other people's machines, we are vulnerable to unwanted, intrusive observation. We see this as a real, bad problem. We fear the invasion of our privacy. Strangers or even people we know may invade our inner zone. That personal, private zone only we can know. It doesn't matter whether we know our visitors, or like them, or fear them, or love them — when they are unwelcome, they are intruders. We know these intruders can get to us through our machines. The strange thing is we have become so dependent on our machines to help us see and understand the world, we depend on our machines to help us see ourselves. We depend on machines to look outside and we depend on machines to look inside. So to be on the safe side, let's begin by making sure our machines are linked only with us. Let's make sure we are off-line when we are looking in, on-line only when we are looking out. Let's try to make sure no one can look in when we are looking out.

[2] The Man Wore Dark Glasses, Chewing Gum But Saying Nothing.

Wearing headphones with his music. Just watching. Not interacting, overtly. Hands in his pockets, he scans the room. He plays the games at home, where nobody knows the score. He plays with his machines. He never dials out. His machines stand alone. They act like him. They don't have a choice. They are under his control. Machines are great that way. They don't want to make choices. Freedom is just another word for out of control. The nice thing about being alone with a machine is that sense of consistency. When you are looking at yourself through a machine, you're pretty sure you're going to get the same picture, night after night. When you are home alone, inside yourself, consistency is very important. Mental routines are cast in precision steel. Tight circles. Mirror upon mirror. Oh, the range of experience in my apartment. I'm completely free to be who I am. I'll never let you in. My machines are all the company I need. I can relate to them in confidence. They are reliable. No surprises. No intrusions. No others. No regrets.

[3] Just Because the Things Inside Can't Be Measured, It Doesn't Mean They're Not Real.

I was asked the other day to tell someone new who I was. I did. It was a primitive description. I remember thinking while I was speaking that I was selling myself short. I know myself better than anyone else, but I carry so little of what I know to the surface—to the place where the words are. My self-consciousness. Most of what I'm aware of comes my way when I interact with my machines. Interact isn't exactly the word. I use my machines to find out about myself. Some people like the solitude of nature—they like to sit on a rock in the middle of nowhere, where it's quiet and still and they can become one with a lake or a sky. You know they talk with their minds to trees and clouds and little brown birds. I talk to my machines with my mind, my voice, my hands, my fingers. I like to use my machines to help me remember things. And then I like to play back things I remember. I like to re-remember. Sometimes my machines capture things I didn't even notice. I forget the details. But my machines help me collect and assemble details and images and music and things that are important to me. Sometimes I don't know why I collect certain things. I just work with my machines, and there's a moment every once in a while ... A moment when a machine tells me something about myself. Something I wouldn't want to share right away with someone else.

[4] When I Go Outside I Never Travel Alone.

I move with one of my machines, and I reach as far and as fast as it will take me. We're tight and we walk in step. When people encounter me, they see us. They get the package. We're inseparable. If they try to pry us apart, we've learned to hold on tight. Before it gets to the point where my image starts to break up, we'll send out a few adjustments. This is the way I really talk. This is the way I really stalk. I only come out for a fix. I come out to feed. My machines help me find nourishment. We feed together like little brown birds. Foraging in malls. The people without machines are out back, dumpster-diving. We are much more sophisticated. We order frozen meat with remote controls. We watch the meat channel and push the button when we want to make a kill. But that's just my self-image projected into the chill of a freezer compartment. Look out, I'm on a mission to find something to feed on. I've got to find a free port. I've got to dock. I've got to make decisions. I've got to fill the available memory. I've got to release before I'm recognized. And I've got to cover my trail with the moans and groans of machine love, machine hate, machine fear, machine void. Machine love, machine hate, machine fear, machine void. Before you can say interactive carnivore, we're back in the protective shelter of our home.

[5] I'm a Personal Human.

In this world of machines, where people are one with systems and World Wide Webs and instinctive behaviour is valued because it's natural, I'm a personal human. I'm special. Not too unique but relatively insubordinate when it comes to clock time and anticipatory sets. But then again, I fit the mould. I'm a personal human who stays well within the limits of the systems I play. I act instinctively because logic and reason are best left to the machines—just like I don't try to remember details any longer than I have to. I re-remember a story about spiders weaving their webs in outer space. When they first started doing experiments aboard the shuttle, scientists sent a couple of orb weaver spiders into orbit. They wanted to see if these spiders could weave their elaborate webs in zero gravity. The astronauts set up a movie camera to document the spiders weaving their webs. Due to their very busy schedule the astronauts couldn't wait until the spiders started weaving to start filming. They left the camera running for a couple of hours and hoped for the best. When the camera had spent its film, they checked the spiders' compartment, but there were no webs to be found. A few days later, when the shuttle was back on earth and unloaded, scientists found two finished webs, which had apparently been constructed in zero gravity. They were distorted, but true to their elaborate form. Instinct had prevailed in a strange and hostile environment. The orb weavers had managed to act naturally in the most unnatural circumstances. When we forget what it's like to act naturally, it's hard to know when to start the camera rolling.

[6] Memory Has Replaced Logic in Today's World.

People can't carry an argument beyond their position of self-interest. We believe only what we remember to be true. We remember little on our own. We rely on our machines to substantiate the past. Machines have a facility for memory that is precise and extensive. Machines free us from the responsibility of storing and organizing memories. When we forget—and we forget—we simply have to search our machine memories to re-establish what happened. With the aid of our machines, we don't just remember, we re-remember. And in partnership with our machines, we establish the truth by comparing records, the documentary evidence of the past. We establish our personal perspective by re-remembering. By re-re-re-remembering . . . By re-re-re-re-re-re-re-remembering. We stack up the records, the weight of documentary evidence, against the present, and we get a sense of our personal perspective. We enjoy the consistency of our perspective. We are anchored by our memory, our machine memory. We depend on our machine memory. We are lost without our machine memory. When our systems crash, destroying all or part of our recorded memories, we have to decide to rebuild or to pull the plug.

[7] Interactivity and Identity Are an Interesting Couple.

After all, everything I'm saying is hinged on relationships, the existence of relationships. People want to interact, to be a player, to be a player in a game that suits them. When people find themselves in a relationship with a machine, a machine that they control, interactivity is a given. Machines that remember well must be driven. They have to be turned on and turned off. They have to be looked at and listened to, touched and spoken to. They have to be fed. They have to be controlled. If they are to be part and parcel of our inner self, they must be secured. We must be able to expand our internal domain without the possibility of a leak. Secrets are hard to keep these days. Information is flying around at light speed. The world outside ourselves is loud with a dense, opaque fog of messages. Other people's messages. Other people's secrets, amplified and petrified and compressed and homogenized. Other people's secrets become my secrets. It's hard to recognize my own secrets sometimes. I use my machines to focus my attention. I use my machines to lower the volume, to turn down the intensity, to identify a field of secrets that are mine — or at least I think they're mine — I identify a field of secrets so I can relate, manipulate, and operate . . . effectively.

[8] My Machines Are Patient with Me When I'm Having Trouble Finding Myself.

Inhumanly patient. I can take my own sweet time determining who I am. I can build a model of my inner self. I can interact with this model, have a dialogue with my twin, feel progressive as I come into focus. I can feel like

I am somebody. I can feel strong enough to go outside. Not as a bundle of everyone else's secrets, but as a creature with internal integrity. Once the model self is constructed to a point of satisfaction, in both complexity and relative clarity, it is time to enjoy my own company. Until I get bored. First, there is joy, and then there is disappointment. Early recognition and appreciation give way to predictability and numbing redundancy. Arousal diminishes. Even when the loose ends begin to unravel, there's not enough change to generate surprises. At this point, a self-destructive act is prescribed by experience. Erasing memory is one option. Full disclosure is more extreme. The modelled self demands to be jettisoned. Permission is given. You instruct your machine to open its channels to the world outside. You transmit and you submit to intrusion. At the instant of full disclosure, there is total silence. Empty wonder. Is anybody out there?

[9] Why Do I Call My Personal Technological Devices Machines?

I am interested in the way we create personal identity through the use of objects. When you look around others' living spaces and imagine who they are, whether they are neat or messy, nostalgic or romantic, creepy, austere, trendy, or basically unconscious, it is easy enough to imagine what goes on inside a particular someone's mind. When someone invites us into his or her home, we learn a lot by simply experiencing the ambience of the living space. Domestic architecture is like a machine for reinforcing and transmitting a deeper, hidden interior space. The inside of a home is constructed to permit its owner to feel comfortable and natural and to transmit these feelings to others—one's family, friends, and acquaintances. Personal technological devices flesh out one's inner space, whether it's a microwave, a vibrating, reclining chair, a home-video theatre, a police-band scanner, or a personal computer complete with a multilingual voice-recognition package. The task of this machine called home is to permit one to be oneself and to reveal who this person is to others. Most interior spaces are neutralized by self-consciousness and strategic efforts at impression management. Superficial as most interiors are, the personal technological devices are the most revealing objects in anyone's home—these machines are loaded with personal information. If we can find out what is in the oven, or what channel the TV is set to, when or what the computer is used for, we can enter our host's inner sanctum. Getting into the voice diary of someone else's personal computer rivals even the most complete investigation of a stranger's medicine cabinet.

[10] We Are What We See When We See It, Whenever.

A number of people are telling me that perception is everything. Most of them mean they are concerned with how they are perceived. A far smaller number is interested in how they perceive others and the world we live in. Machines can play an important role when it comes to shaping and refining perception. Perception is commonly a half-conscious thing, where you find you are not really paying strict attention and then all of a sudden you find your mind is saturated with your immediate environment. Microseconds later, you become aware of your complete immersion. This awareness is what we call experience. As we become conscious of our experience we find the need to describe what we sense. Our machines are good at description. As we successfully automate the act of description, we increase our capacity and thirst for perception. We create an engine for generating experience and storing it and exchanging it and distributing it. We distribute perception and experience through machine-assisted description: you are what you see when you see it, whenever. Again and again and again.

[11] I've Done a Lot of Flying in My Lifetime.

After a relatively slow start, I've spent a ton of time in the air, commuting mostly on domestic flights, choosing between peanuts or cookies, and steak, fish, or chicken, and trying to cope with the effects of pressurized cabins on my inner ear. I have absolutely no fear of flying. I prefer to sit over the wing in an aisle seat. I disappear into a book or a magazine or a newspaper, and I pretend the plane never leaves the ground. I ignore the safety demonstrations.

I never use the toilet. I never talk with my fellow passengers. This morning a strange thing happened. I put my coat and briefcase in the overhead storage compartment and sat down and buckled my seat belt and put on my stereo headphones and looked around. I knew every single person on the plane. I had met everyone on board at some point or another over the years. Without making eye contact, I identified every boarding passenger, and then I stood up and fiddled with my briefcase in the overhead compartment so I could double-check everyone seated behind me. I knew every single fucking person on the plane, not by name—but I remembered things about all of them. I was afraid I'd be recognized, so I hid in my newspaper and thought about how strange it was that I knew everyone on this flight. That's what happens when you fly around a lot, for a number of years. You find you know everyone just a little bit—and in most cases, a little bit is more than enough.

[12] I Used to Know This Man; He Was My Friend.

I used to love to talk with him. We'd never get tired of talking and laughing. We'd drink and smoke and talk about everything and everyone. We were kind of wild. We'd do stupid things. But time has passed and we've changed, and now we act differently when we get together. I guess we've both changed, but I feel like I still want to talk and laugh and be crazy, yet there's no room. I feel my friend doesn't like me anymore and that he's disappointed with the way I've turned out. We've known each other for a long time. We still see the world in a similar way, but he has lost his sense of humour. He seems to carry a kind of terminal sadness with him. He can no longer experience joy. Everything around him gets flattened by the weight of his attitude. He went through a bad run, where everything went wrong. A string of things didn't work out. The disappointments piled up over time to where he couldn't dig himself out. He stopped believing things would get better. Finally, he succumbed to the rhythm of falling—of coming up short. Even when there was a clear break in the negative pattern, when the overcast sky opened up to the heavens, he couldn't lift his mood. He was locked into a miserable self-banter. He had lost his sense of humour. I hope I'll never lose mine.

[13] A Way of Being Alone without Being Alone.

I like working with my machines because it's like being alone without being alone. Machines are part of an ecology of human nature. When we are one with a machine, we are connected with all machines and everyone else who touches machines. We are one with human nature, and it feels comforting to be part of an ecology of people connected by machines and machines connected by people. The thing I like about machines is the detached intimacy I can establish and maintain between myself and others. I can be alone with someone else. I can leave my body and be close without touching flesh. Flesh to flesh is one thing; being alone with someone else without touching is another. Human nature is tied together with machines of all varieties. It's a world wrapped around a world of other species and other times. It's a point of view that depends on a shared solitude, a pleasure derived from being alone with others.

[14] You Can't Teach an Old Machine New Tricks.

You can make a new machine do ordinary, dated things, but why bother? Most people in love with machines are just going through the motions. It's not important what they do; it's important how they feel. You can see this indecisiveness in people who love to watch television. People love television because they enjoy being alone with hundreds of thousands and even millions of other people. Simul-taneously. They sit and watch popular programs, and they feel part of something bigger than themselves. When the phone rings, they expect to receive a personal, exclusive message from another individual. They are not interested in having a television experience when they answer the phone. Of course, video phones will be fair game for advertisers and infomercials. And in the flip-flop of a machine-identity crisis, our television systems will become a new territory for personal, exclusive messages. Video and computers are already being used like telephones, for personal messages, but for a lot of people that's not very useful. These new interactive machines demand our complete attention and direct participation. The problem with interactive technologies is that you can't do anything else at the same time. Parallel activities are ruled out. Listening to the radio or television while reading a book, or watching a movie and daydreaming, or eating and reading a newspaper, or talking and having sex while listening to recorded music, or watching recorded sex while playing live music ... a lot of wonderful parallel activities are being eliminated by interactive technologies. Interactivity, and the exclusionary intimacy it implies, favours memory retrieval in an asynchronous mode. ATM (asynchronous transfer mode) is replacing STM (synchronous transfer mode). STM occurs in shared, real time, and ATM is the end of real time as we know it.

[15] There Is No Longer a Centre for Self-Identity.

The body, no matter how firm, is not the permanent address it used to be. We are all homeless in relative terms. Looking back at the body gives us pleasure. But ultimately, it is just another thing in a world of things that gives us feedback and helps us keep our balance. Constructions of external memory replace the body. We deposit external memory wherever we can. We touch; we revisit these memory constructions. We search for familiar territory. Touching things is good for finding our way. Establishing form through pattern recognition. Establishing and re-establishing order. Checking our references. Making conversations with ourselves. Forming new relationships. Surrogate relationships. Going through the motions. Practising, rehearsing for the real thing. Filling in the details. Embodying stereotypes. Pretending we are sharing secrets. With somebody else. Why is it, when we hold onto our secrets, we feel like outsiders? Why do we feel so disconnected? It doesn't matter. Our machines are very good listeners. They're the last good listeners.

TOM SHERMAN

Memory and Hope (1996–2006) 1996

We inhabit an increasingly complicated technological world. We are willing participants in the creation of a second nature. While the biological world is collapsing, we are busy constructing an elaborate media environment. As a means of describing and coping with this second nature, it is useful to apply biological language to our postbiological cultures. The future of evolution will be cultural, not biological.

Memes, instead of genes, are and will continue to be the building blocks of our evolution. Memes are ideas or concepts (patterns of cognitive behaviour), or simply the way things look or sound or feel. The concept of the meme was launched by Richard Dawkins, an evolutionary biologist at Oxford University. In his book *The Selfish Gene*, he suggests that living organisms are nothing more than survival machines, subservient to their genes. Genes will or will not survive depending on the bodies they construct. The genes are in the driver's seat. We exist to permit them to replicate. They build our bodies, the vehicles for their mobility and reproduction. These bodies are phenotypes, the physical presence of the gene as it interacts with the environment. And these bodies create extended phenotypes: in other species, extended phenotypes are things like birds' nests and beavers' dams; in the case of humans, extended phenotypes include architecture, art, transportation, and communication technology. Extended phenotypes are part of the story of natural selection.

Language springs forth from our bodies as proto-memetic soundings. Memes are cultural building blocks, the message forms and contents at the base of our cultures. Video and film and digital media genres are memes, and they too can be said to share the ultimate goal of replication and survival. We build organizations and networks around these memes to facilitate their propagation and replication. Media production co-ops, exhibition sites, or distribution systems can be seen as survival mechanisms for the memes themselves. The memes drive the phenotypes through languages (protomemes) into extended phenotypes, the cultural infrastructure of the organizations and the networks.

Memes are also the basic units of cultural inheritance. They are our thoughts, ideas, images, and sounds, carried by our appearances and voices. Emerging from our meme pools, dominant memes are driving our phenotypes into a state of hyperphenotopia. Our extended phenotypes are becoming more and more extensive and interwoven. The continuous evolution of our extended phenotypical environment becomes particularly apparent in emerging technologies.

Emerging technologies are the new "species" of extended phenotypes. The development of the phenotypical environment can be seen as a survival strategy for the dominant memes of our times. Obsolete technologies falter and fall away as emerging technologies move in and spread. Memes are altered as they are forced to compete with other memes in the cultural environment. They adapt, or they disappear.

The proto-memetic energy of language stirs the meme pools. Memory is the essence of who we are: the present and past combined. Memory is order and certainty. It is controllable through revision. Hope is the future, the unknown, the dream, and/or the nightmare. We hope to envision what the future will bring. Our extended phenotypes are right in front of us. Emerging technologies point the way to the future.

Why are we so motivated to find out what tomorrow will bring? Most technology futures are in the stock markets. People play the future hoping to make money. If money's not your game, you may simply want to make sure you're not outflanked, exploited, and destroyed. Or more optimistically, you may wish to seek ways to fulfill your memetic potential as a person-machine entity. Machines R Us. People look to the future so they may seize opportunities as they arise.

Predicting the Future

There are a number of ways of predicting the future. None of them is foolproof. One good rule of thumb is to look back thirty years for every ten you try to look ahead. Time is a continuum. If we drop back to 1966 and trace a wiggly line to the present, then we should be able to extend that line ten years into the future, right around the corner, to the year 2006.

In 1966, we were living with minimal cable TV and wobbly satellites. Computers were strictly for calculating, but there were already lots of cartoons about thinking machines. By 1969, the U.S. Defense Department had set up its own computer net, ARPA Net. Sony introduced Betamax video in 1972, and VHS emerged shortly after. In 1980, Turner fired up CNN, the first twenty-four-hour news net. In the early 1980s, IBM introduced its first personal computer (1981) and the U.S. government created NSFnet (National Science Foundation Net, 1983), the germ of the internet. In 1989, the World Wide Web was created in Geneva, Switzerland. In 1993, Bell Core transmitted full-motion video over standard twisted-pair telephone wires.

Artists during this stretch were busy, desperately engineering their own kind of memes, or at least they were trying to serve, conserve, and preserve the interests and integrity of waves of naturally emerging alternative memes. After a decade of abstract expressionism by alcoholic men, advertising around 1963 spilled over into fine art with Pop Art, and then there was one last gasp of modernism, an attempt to separate art from the world at large again: minimalism and conceptual art.

Analog to Digital

By 1969 to 1972, video art was finding life support on a growing equipment base, and it rose quickly with the energy of the sixties and the desire for peace, along with environmental and feminist concerns. By the mid-1970s, women made up half of the video and performance artists in the Western world. Then in the 1980s, art became all tangled up in entertainment, and things were so fucked up for a while that painting as a real-estate business made a comeback. First, artists were critical of entertainment, wanting only to inhabit its seductive forms to attract and subvert an audience. But, unfortunately, many artists became indistinguishable from entertainers. This attempt to entertain was a strategy for survival, but it left most artists beached and endangered in the 1990s. In the information era, information is a commodity and art is linked with entertainment. In the emerging information economy, art has been pigeonholed as nothing more than difficult entertainment.

As an artist, I hitched a ride on the emergence of video and couldn't help but notice the digital rhetoric rising to the surface between 1972 and 1974. I remember how all the shortcomings of analog video were addressed by a then-inevitable but still inconceivable digital future. There was a lot of speculation about analog-to-digital conversion, mostly to improve editing and dubbing, and therefore the reproduction, or perhaps more correctly the replication, of our messages.

The sociopolitical drive during the 1970s was to gain access to the means of production, distribution and exhibition: access to equipment, distribution systems, and exhibition sites (that is, audiences). Our memes, which carried different values than the predominant memes of mass or popular culture, could not survive without a contemporary media environment of extended phenotypes. Meanwhile, our societies' predominant array of extended phenotypes were transforming from analog to digital.

Real-Time Digital

By the end of the seventies, a minority of artists had made the compromises necessary to gain access to digital technologies. They had become either researchers or mascots (creative types, the fashionably hip) for governments or corporations. They had gained access to the digital technologies that only governments and corporations could provide. In 1979, I was working on a contract with TV Ontario in Toronto, conducting research for a television series called "Fast Forward," which dealt with how the emergence of digital technologies would affect various sectors of society. My contract was for the development of the episode on the arts. After we talked to hundreds of people on the phone and examined the videos they sent us, the program ended up featuring Steina and Woody Vasulka, Kit Galloway and Sheri Rabinowitz, Richard Lowenberg, and Laurie Anderson, among many others. The buzzwords of the day were *real-time digital*. The visual artists envied the musicians and composers because digital audio, with its relatively modest bandwidth, could already be made and shared in real time. The hope was for real-time digital multimedia including a *live* visual, moving-image component.

From 1981 to 1987, I served as video officer and head of media arts at the Canada Council for the Arts, Canada's national arts funding agency. We launched the Computer-Integrated Media grant program in 1983, anticipating the importance of the digital arts. In 1986, I was, with co-commissioners Roy Ascott, Don Foresta, and Tommaso Trini, involved in curating digital works for an international exhibition, *Art, Technology and Informatics,* at the Venice Biennale. This show featured an international network of communications artists who had worked collaboratively in global networking as an end in itself for years. Real-time digital had arrived, somewhat lamely in the form of low-resolution, slow-scan video, decent audio (over standard phone lines), and shared-screen text and drawing experiments. The whole Venice exhibition was organized and held together by an e-mail network based on a commercial business network called the I.P. Sharp Associates system that had been used extensively by telecom artists since the early eighties before widespread use of the Internet.

To get a perspective on how much has happened since this Venice show in 1986, fax machines were still relatively rare during this period. Artists were still going to a lot of trouble organizing collaborative city-to-city fax-

exchange shows using donated, then still expensive, fax machines. This same year, the price of fax machines plunged, and there was an astonishingly rapid assimilation of telefacsimile technology throughout the telephonic world.

By the end of the 1980s, there had been major advances in personal computing and networking and remote sensing, optical memory, robotics, smart systems, and artificial intelligence. There was an explosion in the complexity and power of our extended phenotypes. Primitive forms of artificial life were beginning to emerge (the first digital codes as life forms). Although this time was a period of exotic technological change, artists and arts organizations were becoming interested in ordinary things, like computer databases. Building databases appeared pedestrian and hopeless, but this job was smarter work than it seemed at the time. Only those who could get their data organized would be prepared for the browsing public. To restate the obvious: art will become difficult entertainment in the information era.

Other artists were obsessed with inventing new technology to suit their needs. These were the hands-on inventor types. Digital-tool making was and still is a popular preoccupation with several artists. Process has replaced product, except where tools are concerned. Device development is an R & D role for creative entrepreneurial types. There are big bucks in tools: hardware, software, and all future in-between-ware. Sophisticated multimedia emerged, integrating light and sound, dance and music and images, canned and live images simultaneously. Multimedia linked the statistical and anecdotal, and the virtual and real. Unfortunately, the technologies reinforce, if not create, a class system where the tool makers are more often thought of as professionals and the tool users as amateurs.

Video Dial Tone

This account takes us up to the present, so now we can look out into the future. I've organized my thinking on the near future by borrowing heavily from John Pavlik's *New Media and the Information Superhighway*. I have of course distorted his ideas to suit my more intuitive approach. We are still chasing the dream of real-time digital, but we are getting closer and closer. While we continue to spend too much time waiting for files to download and unpack, the gap is closing. Digital media are beginning to behave like analog media. The next big breakthrough will be *video dial tone*. Eventually, we'll just order movies at home using computers connected by phone lines. Movies will be digitized and loaded onto video servers, and true video-on-demand will be a reality. This capacity is one major part of the information market the Baby Bells want to capture. To make video-on-demand work, they will have to offer full control to the user: play, fast forward, rewind, stop, pause. The delivery of fully interactive video, the Baby Bell executives tell us, will usher in the era of the twenty-five-cent movies (talk is cheap while the systems are still just hype). But the message is: Do not buy stock in chains of video-movie rental stores.

Coping with Abundance

With corporations the scale and influence of Microsoft, AT&T, and AOL involved in fusing telephone, broadcasting, publishing, and cable into one industry, twenty-five-cent movies will eventually compete with fifteen-cent stock-market tips and five-cent glimpses of our favourite body types. The main challenge for the information consumer is and will continue to be coping with abundance. There will be a need for personal informational services, navigational, and preference-enforcing tools. Personal digital appliances, multiple generations beyond the Apple Newton, will become important survival tools.

Besides just coping with abundance and managing to shovel through all the crap to find useful information, what emerging technologies will be useful for independent, alternative media producers in the near future? The answer is that all extended phenotypes are and will be potentially important for memetic replication and survival. Just survey the media environment to find a place to play. Look at what's happening on desktops alone! Desktop publishing. Desktop video. The desktop darkroom, sound studio, radio station, video performance cabaret. The virtual city and countryside on a desktop. Network your desktops.

In coping with abundance there is a lot of thought and programming going into intelligent agents, personal artificially intelligent assistants that learn by observation, make decisions, filter incoming information, and select and model data for their owners. Artists could devise agents that perceive and conceptualize life as art. Artists will create intelligent agents that control and destroy stereotypical traps and force their owners to stick to the high road of prototypical and atypical perceptions, expressions, and lifestyles.

Sensory Dimension Shifts and Perceptual Jacking

Information graphics will continue to be a huge growth area. People will need to be able to *see* patterns and relationships in complex fields of data. Sonification, the depiction of visual information in sonic forms, also has potential, as do 3-D and virtual environments. All kinds of sensory translations, sensory dimension shifts, and perceptual jacking will be useful in refreshing and rehabilitating the numb and burned-out masses in our societies. Artists will have an edge in transposing, transmutating, and transmogrifying (that is, to alter radically with humourous or grotesque effects) data and information.

With the development of computer-network transactional services (home shopping, travel reservations, banking, and bill paying), it is conceivable that all information transactions will be logged and be subjected to billing. Cultural, and even eventually perceptual transactions, will be quantified and metered by time and bandwidth. Art and culture on the Net will produce income for artists when the value of the information becomes apparent.

Encryption, securing the privacy of messages and nature of information consumption, will be a major area of concern. As people are given the option of selecting highly personalized information interactively, personal-info consumption profiles will be revealing. Privacy issues will become paramount, especially as the idea of information moves away from facts and begins to reflect emotional realities. Info-consumers, no matter how demanding, will not want to flaunt their emotional needs and desires. Artists, experienced in surviving the expression of emotional needs and desires, may offer protection for the overly exposed. Aesthetic coding, as a means of obscuring the explicitness of emotional causality, may become far more interesting to a broader public as the demand for privacy and encryption increases.

Emerging Technologies Are New Territories

The above speculation questions how emerging technologies may be used by artists for their survival. As artists, we have our own memes to replicate, and pretending we can ignore the digital revolution is useful only for temporary psychological relief. Emerging technologies are new territories, frontiers, where the quick will inherit the ground, if only to hold it temporarily. Remember, the primary goal is to replicate and adapt to changes in the environment. And to seek new ground when forced to move on. And don't forget endangered, obsolete, and discarded technologies. Old media territories can be revitalized and used to sustain endangered ideas and emotions. Territory is territory in the survival game and technologies seldom become extinct, unlike biological species. Obsolete technologies decrease in monetary value for a time until they become the terrain of traditional art forms or simply devolve into antiques. In other words, obsolescing technologies are also shifting ground, but they're far less volatile and hostile an environment than that of emerging technologies.

Cultural Matchmakers Are Particularly at Risk

Everyone is threatened by the speed of change these days. Artists should count their blessings, given their highly refined sense of identities, finely tuned instincts, and real skills. The professional sector that artists have traditionally depended on is in a far more precarious position these days. The curators, publishers, programmers, and producers—those operating out of institutions and organizations like museums, galleries, artist-run centres, clubs—these middlemen and middlewomen are the threatened roles in a digital era characterized by direct marketing, mail-order distribution, and desktop dissemination. The digital force is ultimately desocializing, and cultural matchmakers are particularly at risk.

Most emerging technologies are attractive because they offer new degrees of interactivity. Interactivity, the immediate control of proximity, is valued highly in cultures obsessed with speed and instant, perhaps superficial intimacy. Is it any surprise that contemplation and analysis are being ghettoized in shrinking intellectual circles? Theorists like Roy Ascott in the U.K. are declaring that the period of analysis of representation is over and done with. The postbiological universe, our second nature in all its glory and horror, is under construction, and those who hesitate to don a construction hat will simply have to inhabit someone else's world. Critical thinking, no matter how threatened, will, of course, continue. To contextualize and define one's position is an adaptive strategy and is essential to social engagement.

The User Is the Content of Any New Medium

In concluding this muse on emerging technologies and their potential for creative people, it is important to understand that the user is always the content of any new medium. Marshall McLuhan said this first, and Microsoft, AT&T, the Baby Bells, Telus, and all the thousands of desktop meme-replicators should keep it in mind whenever they look to create or inhabit new or obsolescing phenotypical territory. Until such time as machines themselves are the primary users of emerging technologies (and, unfortunately, this outcome is not unimaginable), human needs and desires will determine the path of technological development. This forty-year growth cycle of memory and hope can be compressed into one primary objective: to accomplish real-time digital communication and expression. In other words, to accomplish analog realities using binary code.

So far, we have digital technologies for reproducing and permitting more interactive access to reproductions of analogic experiences. Symbolic digital communication and expression, such as text and clearly symbolic image exchange, is possible in real-time digital. Computer music can be created and shared with an audience in real time, but the actual nature of digital music is revealing. When was the last time you experienced satisfying emotional content in a piece of music created with synthesizers or rhythm machines? The missing content in emerging digital technologies is human emotion: outrage, anger, isolation, sadness, frustration, anxiety, affection, love. The missing people in the digital world are emotional people. Their memes do not flourish in the territories of the emerging technologies. Not yet.

I came to this realization one night while I was surfing the Net, the realization that the digital culture, as it exists so far, has an emotional vacuum that far outstrips the content crisis more commonly decried. I have a CD player in my PC, and I was listening to John Coltrane at high volume under headphones as I experienced the Webness of some new, fully dressed multimedia publications on the World Wide Web. I was listening to Coltrane's "Expression," a digitally remastered set of recordings from 1967, toward the end of his life where he takes you through an emotional terrain in twenty minutes that rivals the emotional content of most people's entire lives. This compressed emotion is designed to disturb with its hypersensitivity and sheer intensity and density. The solo titled "Offering" is so powerful it creates an uncomfortably warm zone of emotional memory in my chest every time I listen to it. I can never get over how much Coltrane could feel, let alone how effectively he could share his feelings with me. His sax is an extended phenotype, but the linkage is analog and direct. When was the last time you were moved by art originated in digital media?

Emotional tone will naturally fill all digital forms over the next decade. No matter what permutations of new technologies emerge, emotional people will eventually flesh out sterile digital territories until now inadequate for the exchange and appreciation of normal human emotions.

Nature Is Perverse Sometimes (Version 6.1) 1997

Is our meme pool stagnant, expanding, or collapsing?

If our contemporary cultures are conservative and intolerant of anomaly and change, the meme pool must be stagnant in many regions. If the natural world is being absorbed and completely interwoven with the artificial (or symbolic), then meme mutation is rampant and the meme pool is expanding. If thousands of languages and cultures are being destroyed by a handful of imperialist technocultures, then it is clear that the meme pool is collapsing.

With the advance of electronic and now digital technologies, the balance of culture has clearly been altered and there is change everywhere. Nonelectronic cultures, traditional cultures based in unplugged biological and sociological experience, are endangered and threatened with extinction. As communications and information technologies are introduced to all cultures globally, cultural diversity is reduced. At first, traditional cultures are merely infected with artificial cultures. The initial infection is not fatal. The arrival of artificial cultures is initially greeted as a gift from far away. Artificial, electronic cultures are designed to travel far and wide. They have global mobility, and once a technological infrastructure is established, destruction of traditional, nonelectronic cultures is imminent and swift.

An argument can be made that artificial, even seemingly alien cultures are no less natural than traditional cultures. Man is natural, and everything that man does is natural. If one is in the director's chair of a technoculture, or if one is represented or served by an artificial culture, then it's hard to argue against success. Those not represented or served are quick to be threatened and perhaps smart to resist. There are many who stay away from computers and computer networks simply because they feel awkward and frustrated when in relationships mediated by digital machines.

One strategy to reverse the dominance of artificial, alien, digital cultures is to fill the machines with natural biological and social phenomena, at least symbolically. Whether this strategy uses pictures of animals and plants or vaginas or penises or the megapolyphony of chat lines, the content of any new medium is its users (unless you are excluded—then your exclusion is the content). To illustrate my point, let me introduce some images of nature to this digital text. Viruses and bugs tunnel effectively through elaborate switching systems, waiting patiently for open gates. Weeds and vines twist around and force their way into network backbones. Cockroaches, mice, and rats scurry from Chicago to Melbourne to Lima to Kigali in milliseconds. The little creatures blaze the trails, while people sit transfixed by screen light.

We bathe in the screen light of our computers, sitting as still as plants, facing an artificial light source, its surface articulated with symbolic nature. Artificial life (A-life) and natural life (N-life) are reduced to a barrage of symbols: language itself as the primary experience. When we leave our workstations and go outdoors even nature appears symbolic. The black birds, the crows and ravens, talk to us as they eat our garbage. The squirrels perform high-wire acrobatics as they chase each other from telephone pole to pole across sagging lengths of fibre optic cable. Which species are information rich or information poor? Will A-life and N-life eventually trade places? Will virtual representations of extinct N-species take over cyberspace? Will A-life spill out of our machines and escape into the sunset?

God, I don't know. I do know that we don't appear to give one sweet fuck about the natural environment. We love to model cyberecologies and talk about how representation is a dead issue. It's construction time! A-life is under construction, N-life is under destruction (not deconstruction). The argument is that A-life will help us understand N-life, what's left of it.

Thinking about complex, parallel, interdependent systems (A-life and N-life) is good. Artificial systems can provide insights into natural systems. I can accept all this postbiological, pan-ecological posturing, if it is leading us toward a greater awareness of our global environmental crisis. When I remove my VR headset, I find I'm living in a burned-out, totally out-of-date industrial city in the rust-belt of the American Northeast. Syracuse, New York, sits on the south end of a large "freshwater" lake, Onondaga Lake, which is so polluted it can barely support bacteria, let alone fish.

Why are particular domains of the technological environment so hostile to certain memes?

The seas are empty (once abundant species such as Atlantic cod are threatened with extinction), but the majority of highly educated, computer-networked artists, writers, theorists, critics, and scientists worldwide are more interested in the emergence of A-life than they are concerned about the disappearance of N-life. The memes stating "nature's in trouble," the images of clear-cut forests, waterfowl covered in crude oil, or dead dolphins being removed from commercial fishing nets — these memes have become common in the news media. You will find "nature's in trouble" memes on television news reports and in newspapers and magazines. But you will not find "nature's in trouble" memes on the Net in any great abundance.

This absence of "nature's in trouble" memes on the net can be explained in a number of ways. People living in cyberspace are largely urban creatures; their natural environments have been covered by concrete and asphalt for so long; their air has always been blue-grey with exhaust fumes; they don't count the natural environment when they take stock of their quality of life. Nature, for the urbanite, is *the body*, and the body is thought of as a discrete biological universe for limited sensual awareness and reproduction. For most networked urban intellectuals, the body is the necessary site of the mind; nothing more, nothing less.

In the context of the Memesis project,[1] none of the participants are professional conservation biologists. Thus, it is not our professional business to see extreme, tragic irony when fleshing out the cultural ecology (of memes) during an era of catastrophic biological collapse. There are lengthening lists of endangered, threatened, and extinct species on the Web, but "pictures" of declining air and water quality, of species of plants and animals that have disappeared or are rapidly disappearing, are not on the Net, not yet.

This absence of nature is largely the result of limited bandwidth and fidelity. The Net for most is still an electronic multimedia "magazine" of text, stills, moving picture clips, and fragments of voices and sounds. Real-time digital video of reasonable resolution would offer the potential of better pictures of nature. But then the advertisers would move in. Advertisers would pummel the Net with the "nature is fresh and clean and beautiful" meme attached to and contextualizing whatever they were trying to sell. Anyone with access to commercial television in North America (and far beyond via satellite, microwave, and cable) is inundated with messages about beer that is made from cleaner and colder water in Canada, or land-roving, four-wheel-drive vehicles that can climb mountains, or sanitary napkins that smell like wildflowers in a meadow! The "nature is . . . beautiful" meme is being pumped out into television land in massive doses. The "nature is . . . beautiful" meme also does very well in magazines, billboards, and backlit photomurals.

So, to recapitulate, we find the "nature is in trouble" meme doing well on the television news. We find the "nature is . . . beautiful" meme flourishing in the advertising space of television, billboards, photomurals, and magazines. (And the "nature is amazing" meme is abundant on the cable nature channels; and the "nature is unpredictable" meme is omnipresent on the cable weather channels.) None of these nature memes do particularly well on the Net, not yet.

Advertising designers are engaged in sophisticated cultural or memetic engineering. They propagate and introduce new strains of memes, and they analyze the success or failure of memes or examine new relationships between memes and memes, and memes and audiences. They study and track how audiences adapt to memes, and they quickly alter memes to influence people more effectively. Advertising designers are well versed in the site of memetic reception: human consciousness. They understand thoroughly the social psychology of the advertising audience. Advertising designers are just turning their attention to the Net. It remains to be seen how long it will take them to introduce "nature is . . . beautiful" memes to the Net.

If advertising designers practise memetic engineering on a massive scale, what possible impact do artists have on the memetic environment?

Artists, like advertising designers, have been practising memetic engineering since the dawn of cultural evolution. Artists have been producing somewhat different results with their signals and have been focusing their energy into different pathways. But, strangely enough, memetic engineering procedures employed by artists have yielded a product similar to advertising. Audiences have been able to differentiate between advertising and art memes largely because of previous memetic experience. (Advertising and art are usually received in clearly different contexts.) In other words, people's memories are organized by the kinds of memes they've experienced and stored. Advertisers use the "nature is ... beautiful" meme to instruct people to buy certain products if they wish to connect with nature. For instance, the scenic natural environment is used to sell all-terrain, four-wheel-drive vehicles. If a person has absorbed thousands of hours of "nature is ... beautiful" memes in an advertising environment (such as commercial television), an experience in the natural environment itself may 'instruct' this person to buy a four-wheel-drive vehicle. Nature itself is transformed into a meme.

An artist may construct the "nature is ... beautiful" meme in a similar way to advertising, in an attempt to yield a different result. A traditional approach would be to make a picture of the natural landscape, using oil paint on canvas or colour photography or video in an installation context. The meme in this case is a stereotype: the beautiful, unspoiled natural landscape should still be a worthy subject for art. It is, after all, the birthplace of the human species and therefore should elicit warm feelings. If the artist can send a clear signal to the audience through individual style that this landscape picture is his or her personal vision of the landscape, while proving that the landscape represented is essential and still significant to the viewer as well, then an actual experience in nature may "instruct" a person to buy a painting or photograph or video installation by such and such artist. Of course, the key factor in this scenario is literacy—sufficient prior learning by the audience. If the audience exposed to landscape art is skilled only in absorbing television commercials, and not viewing art, then viewing an artwork that represents the natural landscape may simply prompt the audience to buy an all-terrain, four-wheel-drive vehicle.

The "nature is . . . beautiful" meme can be constructed in any number of ways, but if it is an image of a natural landscape unspoiled by mankind, it will be perceived according to the viewer's prior experience. His or her memory of the natural landscape and/or all memetic representations of the natural landscape will determine how any new landscape experience will be processed. For city dwellers, the natural environment is an abstraction. Four-wheel-drive vehicles are often called pathfinders. They can leave the beaten track and provide access to virgin territory. To find a natural landscape to match the beautiful-nature memes we experience in advertising or art, we'll probably need an all-terrain vehicle. Nothing beats a meme for convenience and symbolic clarity.

Serious artists have pretty much abandoned the unspoiled landscape. Instead, they've embraced a "nature is a tragic background for man's mess" meme, if they are intelligent, aware of their surroundings, and inclined to be critical. Being critical is an important factor in making contemporary art distinguishable from advertising. The advertising designers stay away from the "nature is a tragic background" meme, because it won't sell anything. It's too much like the tragic nature stories on the news, the oil spills, or the Chernobylls. Even "green" products, themselves critical of environmentally incorrect products, are still sold with sunshine, flowers, and the glistening surfaces of clean water.

Artists generally insist on launching their own brand of memes into the world. They can create their own kinds of memes by choosing to twist and pervert stereotypes, reaching larger audiences preconditioned to receive stereotypical memetic products. If they choose to work with stereotypes and thus manipulate audiences predisposed to be responsive only to stereotypes, they pit themselves against advertising designers and the advertising industry's powerful commercial engines of replication and dissemination. In terms

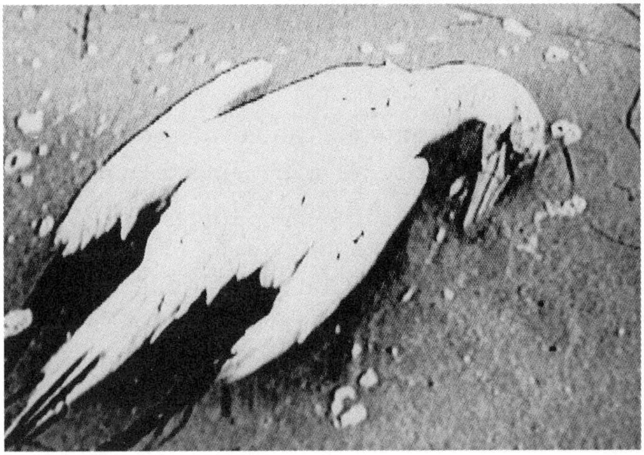

< < <

Dead gannet, as it appeared in Tom Sherman's video, "Gannet Burial," 1999, 4 minutes, and 30 seconds.

of nature memes, this strategy can pose serious problems. It's hard to be effectively critical of those who destroy the natural environment if your "critical" action is to make pretty pictures of the natural environment.

Artists can, of course, play a different, more subversive role in today's memetic universe. They can choose to create and launch anomalous memes, atypical signals, practically unrecognizable messages too personal and obscure to be understood and passed on to many others. If advertising designers and commercially inclined artists push around stereotypes, then it is also apparent that other artists will describe the world with obscure and atypical memes (a-types). Audiences, with the exception of other artists and/or others also conceiving and promoting a-types, will generally overlook atypical memes (a-types).

Up to this time, survival for atypical memes (a-types) and for the artists who issue them has depended on whether these a-types would cluster and emerge as useful prototypes and then finally evolve into stereotypes. A stereotype is an exceedingly common and stable meme, a knowledge structure whose variation and further evolution is compromised because of its usefulness in its current, mature form. Many believe it is impossible to reach a broad audience without dealing with stereotypes. If stereotypes are the raw material at the base of advertising, then artists have a responsibility to work with atypical memes—memes that no advertising designer would touch with a ten-foot pole. Art should exert a force in opposition to or in conflict with advertising.

With the Net, and the horizontal communication and connectivity it engenders, artists and others inclined to create atypical memes must begin to define their own version of nature in cyberspace. Pools of a-types are forming as virtual communities of artists and kindred others meet and exchange atypical images and ideas of nature at various Websites. Unfortunately, the advertising people will arrive with their "nature is ... beautiful" memes sooner than we would like. The pollution of advertising will obscure all other forms of memes. The threat of the advertising industry's imminent arrival should not dissuade us from our own instinctive efforts to bridge the increasing gap between culture and nature. The territory we once knew as the natural environment, no matter how problematic the use of the nature meme has become, cannot be abandoned. Nature cannot be left behind.

1 This text was developed in the context of an Internet-based, international think-tank, in preparation for "Memesis: The Future of Evolution," a live symposium central to the proceedings of the 1996 Ars Electronica festival, held in Linz, Austria, September 2–6, 1996.

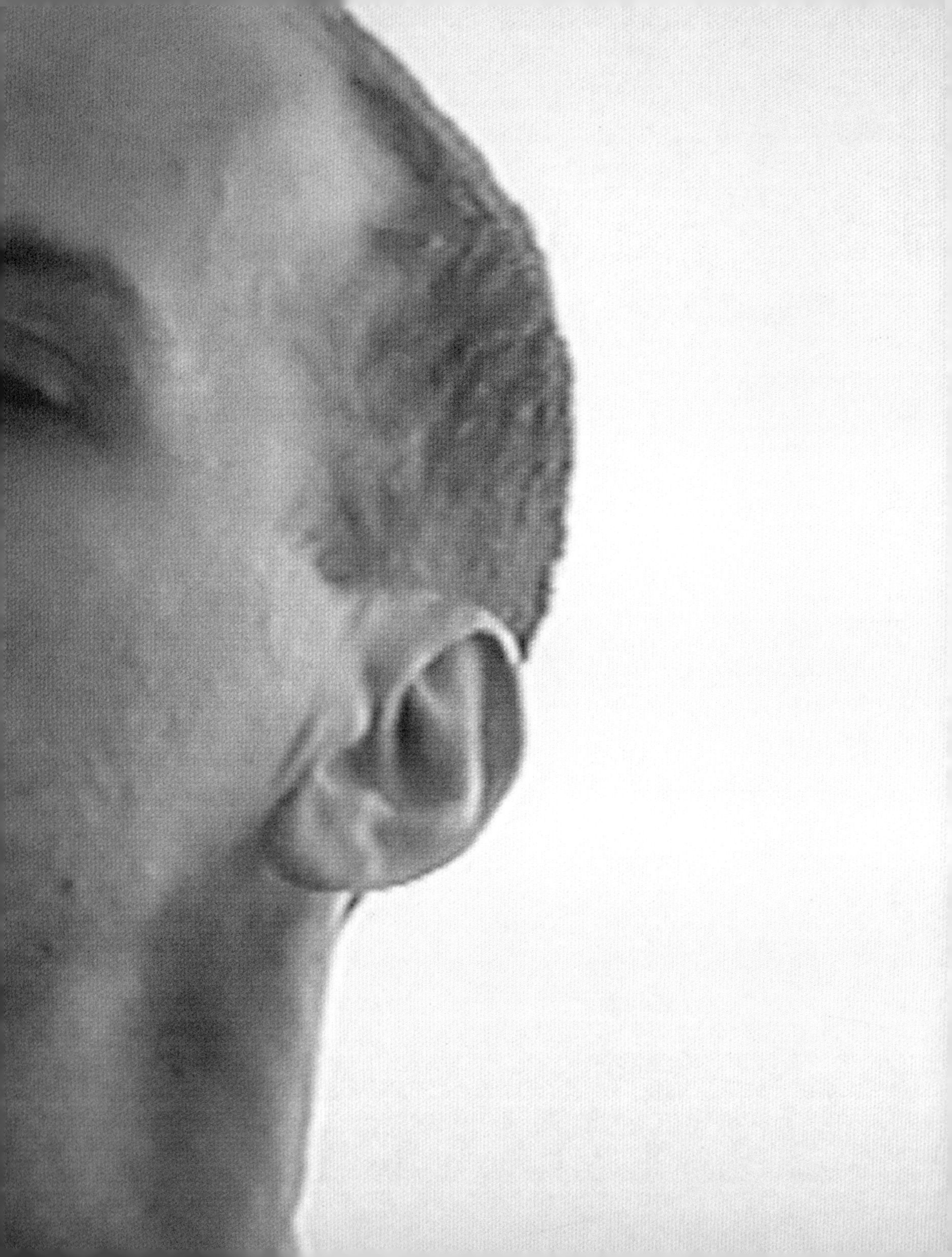

LEARNING FROM OTHERS

Jack's Eye Was a Camera 1991

On a beautiful fall day in 1968, the late Jack Chambers was driving east on the Macdonald-Cartier freeway to Toronto, a comfortable two-hour drive from his home in London, Ontario. He drove a light-green Volkswagen Beetle. He was keeping an eye on the changing southwestern Ontario landscape in the rear-view mirror, as he was in the habit of doing. As he was cruising through the valley marked by the intersection of Highway 59 with the 401, where the 401 crosses through the outskirts of Woodstock, Ontario, the camera in Chambers's mind registered a most profound moment of expo-

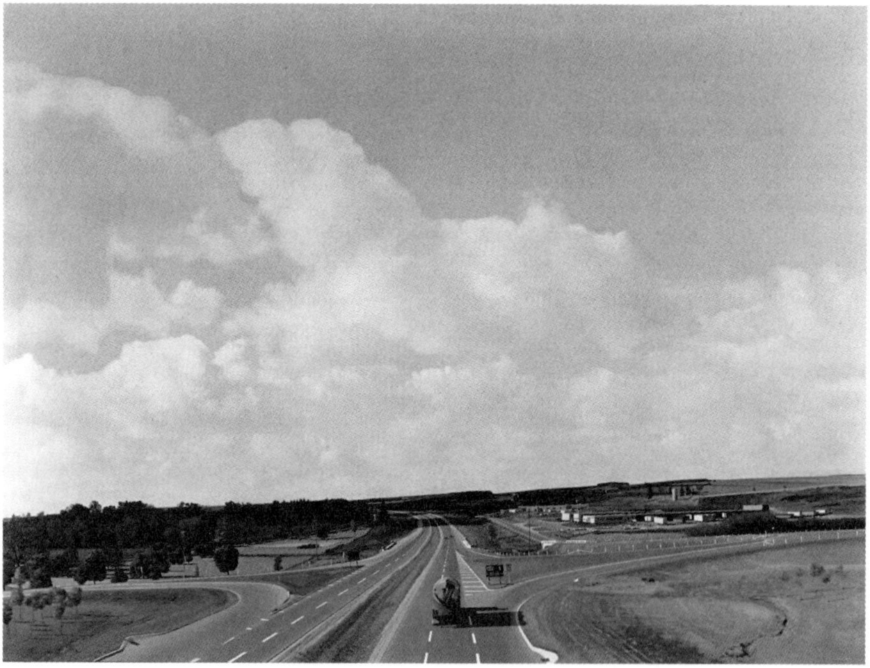

< < <

401 Towards London No. 1 *by Jack Chambers (Canadian, 1931–1978), oil on mahogany, 182.9 x 243.8 cm, 1968–1969. Photo: Art Gallery of Ontario,*

sure. He was struck by the visual perfection of the moment, an instant of perception of his native landscape, recalled as a transcendent event absolutely void of conscious thought. It was as if the landscape had pictured itself.

Chambers, on a return drive to London, sometime later in the same month of October, pulled his car over in exactly the same place, this time facing west, allowing the full experience of this landscape to overwhelm him. He took a colour photograph of the landscape from the Highway 59 overpass, as it crosses over the 401, to describe the valley marked by the 401 stretching toward London. He then made a large painting based on this photograph, *401 Towards London, No. 1*, completing it in 1969. This painting launched a series of photo-realist works, which left his earlier preoccupations with surrealism behind. Few mature artists have ever made such a completely radical, sustained change in the direction of their work in a single medium. From 1969 to 1972, Chambers started and finished a number of compelling photo-realist paintings, governed by his personal phenomenological philosophy, "perceptual realism." During this same period, he also completed three influential experimental films: "Circle" (1968–69), "Hart of London" (1970), and "C.C.C.I." (1971). At the height of this explosion of productivity, Chambers found out he had acute myeloblastic leukemia in the summer of 1969, when he was 39. This disease ended his life in 1978.

Jack Chambers is a mythic figure in Canada. His perceptual realist images form the ground on which much of today's photo-centric art settles. Besides his accomplishments as a painter and filmmaker, he contributed immensely to the notion of professionalism in the visual arts in Canada by founding the Canadian Artists' Representation (CAR), a collective organization for artists, in 1967. CAR established the necessity of benchmark fees, however modest they continue to be, for the exhibition and reproduction of artists' work in Canada. But this pioneering political action will never overshadow his adherence to one fundamental idea which consistently underpins all contemporary Canadian art of merit today. He understood and championed the value of his immediate locale. In a sparsely populated country dominated by European philosophies and American images, he chose to focus on his own landscape and immediate surroundings; and in many of his works, his own family. Through repeated penetrating descriptions of his immediate locale, he was able to transcend its limitations. His discipline and concentration made the ordinary sublime.

Chambers grew up in London, Ontario, a middle-sized Canadian city dominated by conservative British influences. London, Ontario, is arguably the heart of English Canada. It remains conservative with a capital C. Cool, clean, and resistant to change. Creative people have developed there, in spite of the place. In 1961, Chambers returned to London, his hometown, after a long period of study in Spain. By the end of this turbulent decade, he found himself contributing passionately to the "art is life" spirit of those heady days. The sixties were a time for change. He opened his eyes (which at the time spent half of their working hours in communion with the eyepiece of a 16-millimetre movie camera) and found his mind completely saturated with his immediate environment. This he called perception. Microseconds later, he realized how close he was to this landscape and measured this intimacy in degrees of immediate proximity. He called this experience. Description, for this painter-filmmaker, was, in its primary manifestation, a colour photograph taken with a high-quality lens.

Let us take a look at the "photograph" he took of the *401 Towards London.* Our examination of this image is distanced by a series of media translations. We read this "photographic" description through the traditional image-processing technique of oil painting: representational pictorial analogy. We then pass through the print medium of this book, a relatively transparent translation for the privilege of your attention. For those who value the primary experience of viewing a painting, a print reproduction is clearly an analogous representation. A text description of a painting is another form of descriptive analogy. Serial analogies are used to substantiate experience through memory. Memory is a structure of analogies. Such a stack of analogies function in parallel, through a kind of cross-talk. Analogies breed metaphors. The process of sustaining description can be fulfilling.

The landscape depicted in Chambers's *401 Towards London* is the result of glacial forces that created the rich central lowlands of North America. The tilt in the horizon line in this picture, which slopes dramatically to the south, is not the compositional device of a dishevelled camera, but accurately reflects the lay of the land at the site. Although it is not obvious in this picture, the 401 cuts through thousands of square miles of rich farmland. Although the soil of southwest Ontario is not as fertile as that which was pushed farther into the American breadbasket, the dirt surrounding London is as productive as any in Canada. Before the Europeans arrived, the aboriginal peoples of the Tobacco Nation and the tribes that replaced them had deforested the region over time for firewood as they moved west, gradually replacing the tree cover with extensive tobacco and cornfields to sustain their cultures. Tobacco and corn are still the major cash crops today. A silo can be seen just behind the industrial complex to the right of the highway.

This relatively flat landscape offers little resistance to Arctic and Gulf of Mexico air masses, and temperature and humidity tend to fluctuate unpredictably during all seasons. Chambers's picture was made on a fair autumn day with a normal buildup of majestic cumulonimbus cloud at midday. Moisture from the Great Lakes forms such clouds, which roll in from the southwest (bottom left of frame) as the earth turns. As the earth cools at the end of each day, these clouds may darken with the threat of showers, though the pictured clouds bear no real threat of rain and there probably wasn't any rain later this day. A small pool of standing water in the extreme bottom right corner of the frame tells us that there was a heavy rainstorm the night or just days before Chambers took his photo. The highway's drainage system rerouted most of the runoff into Cedar Creek, which passes under the highway just west of the interchange. The dark stand of hardwoods, maple, and oak had benefited from this and all previous runoffs. The fact that these trees still had their leaves would indicate that there had not yet been a heavy frost.

With this atmosphere behind us, let's focus on the highway, its traffic, and its adjacent architecture. This stretch of the Macdonald-Cartier freeway is a modern divided highway constructed in its pictured form in 1955. Southeastern Michigan's twisting, treacherous freeways preceded it, but no highway in the States was more elegant in its reflection of the landscape at the time. A car seemed to drive itself along this relaxing, safe, high-speed superhighway. Remember that Chambers had returned to Ontario from a long sojourn to Europe, where he studied from 1954 to 1959 at the Escuela Central de Bellas Artes de San Fernando in Madrid. He had continued to live and work in Spain until 1961, when he returned home to be with his mother, who was dying of cancer. So this highway was new when he returned, and Chambers must have been impressed with its expanse and utility. Chambers's wide-angle point of view directs our gaze to a point where the highway curves west to meet the horizon, looking out past a truck proceeding toward London. The truck's shadow tells us it is a standard tractor-trailer. The cargo is unidentifiable, covered by a canvas tarp. There are two cars up ahead on this same westbound highway, one passing another, while a single car emerges from a large tree shadow heading toward us in the right lane of the eastbound highway. In 1968, the speed limit was sixty miles per hour for trucks and seventy miles per hour for cars. You could fly by trucks in those days, at seventy-five or eighty. The automobile was king of the road. Now the traffic is seldom this sparse, and both trucks and cars are limited to a hundred kilometres (62.5 miles) per hour. It is the truck's domain today.

To the right of the truck, in the bullnose of an exit ramp, stand two signs, one green and one yellow. The yellow sign dictates a twenty-mile-per-hour speed limit for the exit ramp, but a close examination of the painting itself reveals fictionalized information on the large green sign, probably the only verifiable instance of artistic licence inside the frame of the picture. The sign in Chambers's painting shows a white crown containing the number "59," a white arrow pointing straight ahead down the 401, and the word "WEST." "LONDON," with an arrow pointing to the exit ramp, is on the smaller strip of plywood which hangs below. In fact, in 1968, this sign, which marked this exit ramp for points south, actually read "59 SOUTH" (with an arrow pointing at an angle to the right in the two o'clock position) with the smaller bottom panel reading "DELHI" (arrow in the three o'clock position). Either Chambers couldn't read the sign on his source photograph and reconstructed the sign incorrectly from memory, or he thought the information on the sign would confuse the viewers of his painting, leading them to believe they were looking south to Delhi, Ontario, rather than west toward London. The latter seems more probable to me. Chambers rewrote the sign, making the painting more an evocation of the feeling one has returning to London from Toronto than a magic moment of pure perception framed in the rear-view mirror of a Volkswagen. Londoners prefer their city over Toronto. A Londoner recently told me that when you are driving home from Toronto, and you hit this exact point on the 401, you're exactly half an hour from home.

The site hasn't changed much over the years. At first glance, the basic superstructure of the highway and its immediate landscape haven't changed. There are now telephone and power lines criss-crossing the frame (maybe these lines were present in 1968, but Chambers chose not to paint them?), and the industrial buildings just beyond the aforementioned fictionalized sign have developed into a modest industrial park. When Chambers made his photograph the buildings were the central terminal of Overland Express, a large regional trucking firm. If you look closely you can see the trucks parked at and around the loading bays. There were about fifty cars parked in a lot this side of Overland's building, representing the same number of drivers who worked out of this terminal at the time.

The whole trucking business has been turned inside-out since 1968. Overland Express is now TNT-Overland Express, the result of a 1978 buyout by TNT, the Australian multinational. Such global corporate trends were then compounded by the 1989 Free Trade agreement between Canada and the U.S. This agreement is decimating the Canadian trucking industry today. The Great Lakes trade corridor, the strip of road between Toronto and Chicago, is wide open to North American traffic these days, and American-based trucking firms, enjoying the advantages of lower fuel prices, lower wages, and the advantages of larger-scale operations, are undercutting Canadian trucking. The current Conservative federal government of Canada is obsessed with deregulation and has put a great deal of the transport sector up for grabs in this era of Darwinian economics. Chambers's picture describes a stable physical transport system permanently fixed upon his landscape. At first glance, the industrial features of this landscape appear to be inconsequential and nonthreatening, almost serenely so. But the image, the way it was made and the reasons it was made, functions as a code for the story behind the picture—and the story is changing, advancing, while Chambers's image remains frozen in time.

Chambers intended his paintings to be more than just objects to be seen. He wanted to make objects that could be used literally as devices for seeing the world. He thought of himself as a technologist, and painting for him was a technology. He wanted *401 Towards London* to function as a technological device, the concrete manifestation of individual technique. For such a skilled painter, realist paintings were nearly as transparent a technology as photography. We are asked to look through the "window" of the painting to the horizon. Chambers preferred to think of his perceptual realist paintings as paintings without style. He believed painting styles before the invention of the camera were simply compensations for a lack of visual information. Modern painting, with its reductive path toward the painted surface itself, was anachronistic, as it ignored the camera. Chambers sought to reintegrate the camera with the act of painting in his perceptual realist works, building on a history composed by artists interested in factual accuracy. He was aware that Vermeer, Dürer, Cezanne, and Van Gogh had used grids, light boxes, and later photography to enhance their inadequate visual memories. He showed little interest in the work of other contemporary photo-realist painters. He kept his head down, for although he was a camera man, he was always his own man. Romantically, he wrote about the perpetuation of classicism. He felt it was essential to have absolute mastery over objective representation in drawing, painting, and sculpture. How else could one truly choose abstraction? In the late sixties, in the heyday of conceptual art and during the initial dramatic implosion of modernism, Chambers chose to imitate through painting the experience of taking a picture with a camera. At the time he was immersed in the writing of Maurice Merleau-Ponty, the French phenomenologist. Merleau-Ponty's books, *Phenomenology of Perception, Sense and Nonsense,* and *Signs,* were first available in English translations in 1962–64. Chambers was profoundly influenced by Merleau-Ponty's argument that phenomenological description (for Chambers, the colour photograph) can never be composed of only sensory data, but must also undertake to render the self-transcending or referential character of our experience. The perceptual encounter precedes and differs from (and complements) conceptual thought. Perception is the vehicle by which we gain access to the real.

R. Bruce Elder has written extensively on Chambers and Michael Snow, one of Canada's most eminent contemporary camera-artists. In his book, *Image and Identity: Reflections on Canadian Film and Culture,* Elder argues that Chambers and Snow are the founding fathers of postmodernism, a mindset he sees stemming directly from the film camera and its photochemical reproduction of perceptual experience. Elder's analysis runs contrary to the more prevalent notion that postmodernism resulted from the rupture of history brought on by television and other electronic, instantaneous image forms, which also gave rise to the timely reintroduction of feminine intellect.

Chambers and Snow knew each other through a mutual interest in each others' films. Chambers had seen Snow's film "Wavelength" (1967) at a screening in New York in 1968. Although they were not friends and seldom spoke, Chambers was engaged in a direct, formal dialogue with Snow, as he was with American filmmaker Stan Brakhage, through his own filmmaking. Snow, of course, was then, as he continues to be, intensely preoccupied with the camera's precise capacity for recording experience in photochemical media. When one looks at a painting by Chambers, or a film by Snow or Brakhage, it is almost impossible to avoid self-transcendence and an essential reference to a world that transcends consciousness. All three make "technologies of image" which function as extensions of the camera as technological history. Ironically, all three are perceived today as modernists, as a triumvirate of romantic idealists, which only affirms the confusion permeating this particular age.

I would like to return to the fiction in Chambers's *401 Towards London* because it intrigues me. Chambers chose to imitate through the act of painting what he considered to be the ultimate descriptive process, the taking of a photograph. For Chambers, the photograph was a description of an experience, a moment in space and time characterized by its immediate proximity to pure perception. The act of painting is a manifestation of individual technique, a performance where fiction is introduced to the frame, through either human inadequacies of memory or technique, or the deliberate reconstruction of reality. The frame of this painting is itself a fiction. The image is cropped, not by the aspect ratio dictated by the camera he used, but by Chambers himself. The aspect ratio of this painting is very close to being exactly four units wide by three units high (a rectangle 96 inches wide by 72 inches high). No film camera would yield this frame unaltered. Chambers cropped the photograph in his development and execution of the painting. Rather than speculate on the subconscious influence of the Volkswagen's rear-window frame, I would like to return to my observations about the rewritten, fictionalized road sign. Chambers apparently thought that the truth of the sign captured in his photograph would distort the reality he was doing his best to describe. The information on this sign had to be rewritten to establish a single point of view for the viewer of the painting. "59 South, Delhi" was changed to "59 West, London." After all, the 401, the highway depicted in the painting, is the road to *London* and with this deliberate enhancement of reality, it became, unmistakably, Chambers's road home. The painting of the photograph taken at the site of the experience is a precise, pictorial analogy designed to substantiate the perfection of an exceptional moment—Chambers becoming one with his native landscape.

Canadians suffer when they lose touch with their origins, whether they leave home by choice or due to the forces of urbanization and economic displacement. Regionalism still means something in Canada. Because Canadians are saturated with the images and philosophies of the U.S. and Europe, they long to see their own lives expressed. The fictionalized road sign in the *401 Towards London* says it all. The sign was rewritten to point more clearly to London because it was Jack Chambers's home.

> > >

Personal recollections of Jack Chambers were provided by Stuart Shaw, a lifelong friend of Chambers now living in Toronto; Greg Curnoe, a prominent artist of London, Ontario; and Michael Snow, the artist, musician, and filmmaker in Toronto. I also owe a special thanks to W. J. Brown, traffic analysis supervisor in the Southwest Region, London office of the Ontario Ministry of Transportation. Jeff Brown, through his research and re-photography of the site, verified that the road sign had been fictionalized and provided site information crucial to the development and completion of this text.

REFERENCES:

David Burnett, *Jack Chambers Retrospective*
(London Regional Art Gallery, 1988).

Jack Chambers, "Perceptual Realism," *artscanada*
(Toronto), Vol. XXVI, No. 5, No. 136/137 (Oct. 1969), 7–13.

Jack Chambers, "Perceptualism, Painting and Cinema," *Art and Artists*
(London), No. 7 (Dec. 1972), 28–33.

R. Bruce Elder, *Image and Identity: Reflections on Canadian Film and Culture*
(Wilfrid Laurier Press, 1989).

From a Reservoir of Predictions 1990

In August of 1989, I visited Art Metropole to view video works by artists I thought had done important work during the medium's history. I had been commissioned to select what I felt were exceptional works from AM's video archive, with the objective of pointing out the value of specific video works in historical terms.

Art Metropole's video archive, now part of the National Gallery of Canada's collection, consists of 642 titles by Canadian and international artists. With other video archives in Toronto at Vtape and Trinity Square Video; in Vancouver at the Western Front, Video In, and Vancouver Art Gallery; in Ottawa at the Canada Council's Art Bank, Public Archives Canada, and National Gallery; in Montreal at P.R.I.M. Vidéo and Le Vidéographe; and in Halifax at the Centre for Art Tapes, this important archive shares responsibility for the primary record of Canada's rich history of involvement with the medium. As most video art has been funded with public money, through either direct grants or the support of museums, artist-run centres, and universities or colleges, it is essential that this information be made available to the broadest possible public. (If any broadcast gatekeepers are reading this, the works I've selected are very broadcastable.)

I started my research with a handful of artists in mind, eventually screening about forty works by many different artists. I have always been fascinated with the way artists relate to video—the way they use it to make art—because in the process of their work they set benchmarks for literally *everyone* who will use the medium in the future. And video has quite a future. In 1989, there were an estimated seven million 1/2" and 8-mm video camcorders in use in the world. Camcorder use has increased dramatically since 1989.

I've selected three recordings made by artists when the medium was far less advanced. The recordings I've chosen feature on-camera performances by the artists themselves as they were describing their world through their video. All three recordings offer formal approaches that many others have adopted and expanded upon. The relationships formed and recorded by these artists in video, this art and communications medium, will continue to offer a starting point for those who choose to negotiate or compose their reality through video tomorrow.

The works I have selected are "Sackville I'm Yours" (1972) by Colin Campbell, "Landscape II" (1975) by Les Levine, and "Me$$age to China" (1979) by Susan Britton.

Before turning my attention to these works, let me tell you why I think they demand particular consideration today. All three of these recordings predicted a future that has become an accurate representation of our present reality and an uncanny manifestation of our current concerns. Not only have these artists succeeded in anticipating the reality we now inhabit, I believe their insights will prove to be of great value to future generations as we approach and enter the twenty-first century. These video recordings are precise representations of different perspectives in the seventies that have turned out to have remarkable relevance at the end of the eighties. The relevance of these tapes may be partially attributed to fate, but ultimately the accuracy of their predictions is a result of their individual strategies in recording the exact present they were experiencing in Sackville, New Brunswick; Wayne, New Jersey; and Vancouver, British Columbia. The best strategy for predicting a near future is to strive for an accurate portrayal of the present.[1]

With the above thoughts in mind, I've decided to introduce these tapes in the following way: with a basic fact list about the video work and artist; a detailed description (an analogous text) of each recording, including key verbal statements by each artist during his or her performance on tape; and a description of the context in which these tapes are replayed in Toronto, at Art Metropole, in the fall of 1989.

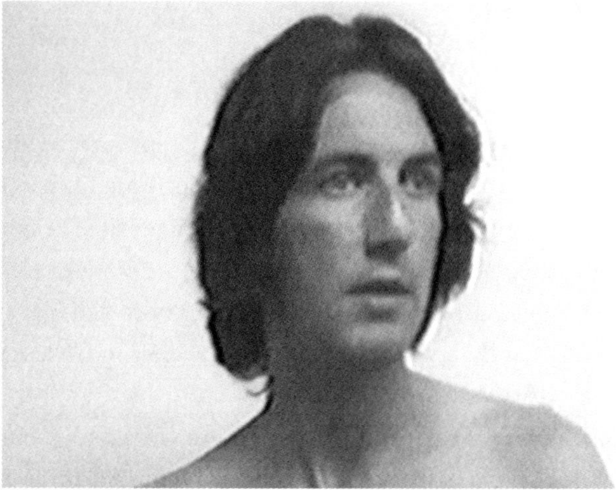

> > >

Colin Campbell, as he appeared in his video "Sackville I'm Yours," 1972, 14 minutes and 20 seconds. Photo: V Tape

> TITLE AND YEAR OF RECORDING: "Sackville I'm Yours," 1972

> ARTIST: Colin Campbell

> BIRTHPLACE AND DATE OF BIRTH: Reston, Manitoba, 1942

> CITIZENSHIP AND CURRENT RESIDENCE: Canadian, Toronto, Ontario

> LENGTH OF RECORDING: 14 minutes, 20 seconds

> VIDEO FORMAT AND FORMAL CHARACTERISTICS: 1/2" reel-to-reel, black and white, single camera, continuous (real-time) recording of an improvisational monologue; mono sound from single microphone

> PLACE OF RECORDING: Mount Allison University, Sackville, New Brunswick

"Sackville I'm Yours" opens in video "black," with Colin Campbell's voice: "No. I couldn't do any more interviews today." Mid-sentence the camera iris opens up to an image of Campbell, shirtless, in a close-up, cropped just below the collarbone. This recording consists of one shot, fourteen minutes and twenty seconds long. This tape is a perfectly framed document of a video performance. Stylishly minimal at its time of creation, there are no frills. As the performance begins, Campbell tries to brush off a request for an interview from a reporter. The reporter is persistent. Campbell tries to set up an appointment for a later meeting. The reporter again wins out. Campbell agrees to do an immediate interview. The reporter is imaginary. There is no reporter. This is a rehearsal for an interview. It is Campbell interviewing himself, a self-interview. He is performing for the camera alone, but there is a camera operator, someone who manually opened up the iris, who remains somewhere in the room. Campbell plays to the exact location of the silent, off-camera interviewer, off to his left, slightly behind the camera. His play to the interviewer off to the side is formal both in the convention of the interview and in its function of setting a photogenic angle to his face, which is rich in soft shadows. Campbell's face transmits an undulating range of truth and fiction in a stream of facetious statements about living and working as an artist in Sackville, New Brunswick. His voice delivers a clean, compact oral essay on the psychology of the artist looking out from a dead-end situation. This tape is a goodbye note to Sackville and everyone who lives there.

Campbell lights a cigarette, which he holds and smokes lightly for the length of the performance. His self-interview goes like this: "No, I couldn't do any more interviews today. No, I can't see my way clear ... My name? I thought everybody already knew. Art Star. Living and working in Sackville, New Brunswick. Sackville is a great little town. What does one do in Sackville? Many things. I can say frankly, there's nothing I haven't done or seen in this town. For instance, I've been to the dump. It's a great little dump. It's probably one of the very best dumps [in the country]."

The timing of Campbell's delivery is halting in an elegant, poetic way. He suspends his words mid-sentence against a series of changes of facial expression. "I've had a tunafish casserole at the president's house. The president of the university [Mount Allison University]. The minority groups, we get together. We spend evenings talking. Well, I'm the only Art Star. I'm the only one listed in the phone book. It won't take you long to check." His wife had once phoned everyone from N to Z during a provincial election in half an hour.

"Well sure I have friends, minority friends. One of my best friends is Art Gallery [Chris Youngs]. Art, he's the custodian of the Owens Art Gallery. Yes, Art and I are friends. We talk."

Then there is his "very, very good friend" Linda Trentini. "She's a racial minority. We can't all be celebrities. She's Italian. I'm not really sure if she is Italian really. She's married to an Italian. She's very interested in my work. Of course, everyone in Sackville is. I have a tremendous response, almost down to the last child, to my work. It's very gratifying. Of course, that's one thing that keeps me here, is this 'go to it Art Star' attitude. Everyone waits for my latest work almost breathlessly. It's very difficult to leave Sackville." Here Campbell begins to talk about his contacts outside Sackville. He traces a string of cities from Cornerbrook to Vancouver (not mentioning Toronto), right down to the West Coast of the States. He ends up in California and Los Angeles. His drifting voice runs through several poetic transitions from the general to the specific: "Los Angeles, yes, Ann, Annie. Yeah, Annie Brodzky and I, of course we've known each other for some time." Brodzky was the editor of *artscanada*, Canada's only national arts magazine at the time.

Campbell sketches out a minimal Canadian art scene, the way it was then. He talks about his other "good friends," including "Dorothy" (Dorothy Cameron, the infamous Toronto-based curator with the great eye) and Art Bank (Luke Rombout, who used to live in Sackville, but left town to start the Canada Council's Art Bank in Ottawa). It was easy to identify who the right people to know were in 1972. All of these important friendships are, as Campbell admits, circumstantial and superficial at best. He carefully frames a bouquet of white lies for the wonderful, warm little town he is leaving. He is leaving Sackville, all of its good points aside. He will miss his special

parking privileges. "Well, I can park my car anywhere on campus for short periods, and anywhere in town." He will miss the fine food and the "quite nice drinks" at the Marshland's Inn. He will miss the "timeless quality" of the place. "The countryside, the ocean, it's just so timeless."

Campbell moved to Toronto in 1973, less than a year after "Sackville I'm Yours" was made. In Toronto, as he had speculated in front of the video camera, you could get a drink without having to buy dinner. You could do whatever you wanted to do, and no one would think too much about it. Sackville had been a severe, restrictive environment. Campbell wanted urban anonymity for his lifestyle, and he wanted a responsive, urban audience for his work. He wanted to know the right people. He wanted to talk to people who understood where he was coming from. Campbell, originally from the Prairies, stationed uncomfortably in the tide-paced serenity of the Maritimes, yearned for the "New York" of Canada. Toronto was calling with its growing skyline and its exciting, young video scene.

At the end of the recording, he checks his wrist for the time. He isn't wearing a watch, but he knows it's time to leave. He puts out his cigarette and ends the interview. He stops speaking and vanishes abruptly with the soft whistle of blank video later dubbed in precisely to cut the performance short. His self-interview ends as it has begun, a minimal document of a video performance. Real-time, one-gun video performance in Sackville. The very beginning of the Toronto narrative style.

Toronto was drawing imported talent in 1972 the way it continues to do today. The difference now is that many more are leaving. As the city has become more difficult for artists to live in, it seems that more are leaving than are coming in. Toronto is on the threshold of another huge growth spurt and organic as these urban population booms may be, it will be described in the *Globe and Mail* simply as the product of economic growth. The money will come from Asia. There are two Canadian cities poised on the Pacific Rim, Vancouver and Toronto. Electronic financial information rules the age.

Artists will continue to migrate to Toronto, the way Campbell did in 1973. It will be tougher landing in some ways. But they will continue to arrive, looking for the same things: urban anonymity, freedom of lifestyle—to meet the right people, to find others who think the same way, to find an audience. Conversely, with the exodus of many others, the regions will benefit. Toronto is now paying the rest of the country back with interest. Campbell's goodbye note to that wonderful little town offers some interesting information for those about to embark into the periphery. Think about what it will be like living in a small town before you pack your bags. This story of the small-town artist on his way to the city is more than just a history lesson. Its viewers may invert its message to prepare themselves for de-urbanization.

Campbell's contribution to the Toronto narrative style is rivaled only by that of Lisa Steele, another Toronto-based artist who also pioneered the literary, first-person narrative in video. Campbell forecast the value of the personal narrative form in video, and over the years he has fleshed it out through a remarkable body of work spanning two decades. With "Sackville I'm Yours," he set the course of his own future. And then he followed his prescription to the letter. Colin Campbell is now virtually synonymous with Toronto, that promising city he has continued to describe so aptly in his own unique voice.

> > >

> TITLE AND YEAR OF RECORDING: "Landscape II," 1975

> ARTIST: Les Levine

> BIRTHPLACE AND DATE OF BIRTH: Dublin, Ireland, 1935

> CITIZENSHIP AND CURRENT RESIDENCE: American, New York, New York

> LENGTH OF RECORDING: 29 minutes

> VIDEO FORMAT AND FORMAL CHARACTERISTICS: 2", colour, four-camera, continuous (real-time) recording of the exterior grounds and interior of a college television training studio, with the artist performing as a television director (video switching, mixing, image processing, and improvisational commentary by the artist, and technical director), mono sound mixed from one exterior and three interior microphones

> PLACE OF RECORDING: William Paterson College, Wayne, New Jersey

"Landscape II" opens with its credits on screen, with the voice of the artist defining the situation for the viewer. Les Levine is performing as a television director in a small training studio at William Paterson College in New

> > >

Les Levine (far right), as he appeared with Joseph Hanf, his technical director, in Levine's video, "Landscape II," 1975, 29 minutes.

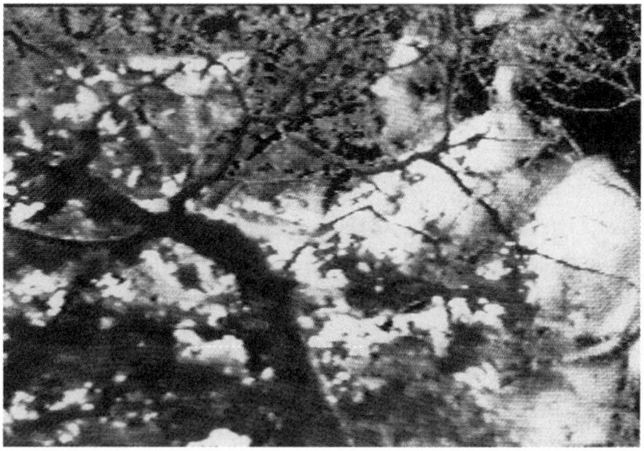

Jersey. He is decisive and confident in the first few seconds, as he is for the rest of the "live" recording. "Okay, videotape rolling. Standby everyone ... Open the show, let's go, first image, take two, take one, take four." The fourth camera offers a surveillance like view of the studio, with Levine and his technical director (T.D.), Joseph Hanf, both sitting in full-body, seated profile at the control console. He asks the interior camera operator (camera four) to tighten up the shot (close in on the directors and the console). The equipment, the VTR, the monitors, and the switcher set the scene. This studio is their link with the college grounds. Three remote cameras provide exterior pictures, a single microphone provides sound from outdoors. A mix of three mics covers the voices of Levine, the T.D., and the studio interior. The bushes, flowers, the ivy on the brick walls, the trees, grass, and weeds, and the sound of birds chirping above the drone of lawn mowers and automobile traffic share the space and time of this recording with Levine, his T.D., and the machines in this studio.

Levine and his T.D. wear headphones with intercom mics attached. Levine is smoking a cigarette in his left hand; his right hand rests on the surface of the control console. After telling the T.D. to superimpose the images of camera one and two into a new "natural-looking" combination of foliage, he directs the T.D. to take it. "Okay, cameramen, you can start looking for other shots

... Focus up there, very good, keep it off the people, this is a landscape tape, we don't want people in it." The cameramen move and re-frame. Levine calls the shots, pausing briefly to ask his T.D. the question at the heart of the exercise: "Should television be used for landscapes? Is it all right to do so?" The T.D. responds: "I think so. It's more interesting than people." Levine talks right through this surprising answer, offering his own premeditated response: "Most people don't think that you should do it, take three. But maybe people will think it's boring if you just have images like this, because they're used to having shows about plays and actors and murders and the news and what have you ... Okay, take what's on the preview here."

Levine taps on the surface of the preview monitor, making contact with his machine link to the immediate outdoors. It is a beautiful summer afternoon to be inside this cramped, modest studio. This studio is brightly lit with fluorescent lights. The artist and T.D. are familiar with each other and work well together. Together, they find a rhythm to switch by. Levine is straightforward and demanding. The work of the T.D. is flawless. They mix shots and witness the immediate results. Levine is anxious to make the best use of the time. He asks for quick cuts to straight, still shot after shot after shot. The straight shots and superimpositions (with dissolves) soon fall to a preference for the look and image-integrating qualities of the video key. For most of the duration of the recording, just over twenty-nine minutes, a matt key technique is used to "paste up" video overlays. The hot areas of flowers (in terms of light) from one camera are "laid over" the cool green leaves of a clump of bushes from another camera and so on. Each keyed image is unique to the video reality; a vivid, artificial, composite image, in video colours true only to themselves. Some of these pictures are beautiful in a traditional aesthetic sense, as framed compositions most commonly seen in nature photography or painting.

Levine asks for a new image and pauses. "That's very nice, isn't it. It's sort of like ... [long pause] ... country." He means it in an amused sense during a relaxed moment of pleasure. Then anxiety permeates his voice again. "Get off that wall, three. This is a landscape tape. Focus on something green

... What have we got here?" The T.D. replies: "Apparently, he's on the side of the building there." Levine quickly: "Get down from the building there. Good, give me a nice green shot, tighten up, good. Let me see a close-up shot, something close with a lot of colour in it." He reprimands the cameramen whenever they violate his notion of landscape throughout the recorded performance. Some of his anxiety is based in formal concerns. Levine: "Camera three, get your camera out of the sky, please. Can you zoom in? Get something very tight. Focus. I want a lot of surface. Take the preview."

"Landscape II" is a video mix of all available foliage—no people, no sky, no walls—no people except for Levine and the T.D., and the machines in the studio. During the recording, Levine does consider including birds, but he concludes that birds are too hard to shoot with heavy TV cameras. The birds are captured in the audio. With their upbeat, high-pitched chirping, they are in constant competition with Levine's voice. Levine and the birds are mixed with machine sounds of high and low frequencies from microphones indoors and out. When there is something on the screen that looks particularly good to Levine, he warms up with genuine wonder: "I didn't realize you can ... Look, you can make landscapes. You can actually make landscapes. It's really so weird ..." Later: "It's interesting how these little blades of grass can be so pretty." "It's sort of extraordinary to look at bits of grass like that, 'cause you hardly ever see them quite like that ... Let me have the dandelion, I want that dandelion in the centre of the screen." The cameraman struggles to centre the weed. The cameramen are the weak link in this production team. They frequently fail to respond satisfactorily to Levine's commands all the way through the tape. They get worse as the tape progresses. Either they are exhausted toward the end or they have lost interest in the shoot. Levine is also more and more challenging with his commands as the shoot progresses. They tease Levine with more shots of brick walls when they are not messing up their shots due to lack of skill.

Toward the end of this experiment Levine calls for slow pans, which he wants to combine in a vertical split-screen. The shots are jerky and uneven. Levine accepts their inadequacies in his professional "on-air" performance as director. He gets the best he can out of his crew. He is far more interested in the look and feel of the mix. This is an exploratory shoot. A selection of well-photographed, well-printed stills from "Landscape II" would be interesting to look at if there were enough of them. There are 183 different images composed and sequenced in this recording (not counting the multiplying factor of the panning cameras, split-screens, or repeated shots due to switching). If all differentiated images are considered, there are approximately 915 images in this twenty-nine-minute recording. During one split-screen combination of slow panning close-ups of two flowering plants, he asks and answers the question of the moment: "What is that like? It's like chocolate boxes."

Levine entered this situation with an open mind. He knew that a string of fragmented images of lush summer foliage would cut and mix into a beautiful brew of electronic landscape. All he asks of his audience is to watch his experiment and consider his fundamental question. Can the second nature of video adequately represent our first nature of plants? The question is posed by an urban intellectual in the anxious space of a television control room. Levine moves through the information very quickly, partly because of his own intelligence and intensity, partly because he does not want the program to be boring. You get the sense there is a lot a stake. Sure, Levine is self-conscious in his performance. He has to get the job done with the means at hand with the presence of the eventual audience looking over his shoulder. But his anxiety has deeper roots. Levine is seeing the landscape in a state of anxiety. Levine's "Landscape II" is an illustration of a state of mind. He is still certain, fourteen years later, that every landscape he inhabits becomes a metaphor for his state of mind. If he can find himself mentally present in any landscape, he is in heaven. It is beautiful and serene in heaven. If he fails to become part of a landscape psychologically, he finds himself in hell. This kind of hell on earth includes the terror of dislocation, which also manifests itself as denial, or more conveniently as boredom.

In Toronto, you can check into the media landscape through cable television. There are about thirty-five channels on cable these days. If you roam all the available channels, you will find three or four channels of natural wildlife at any time of the broadcast day. Amid the game shows and talk shows and movies and news, you will find ducks swimming across the surface of a generic pond, or fish showing off their colours in a sea turned into an aquarium by your screen, or any number of other wild animals performing their natural duties under the reassuring presence of the voice of a relaxed narrator.

Most of these voice-overs are men with deep voices. Women are still rarely chosen to do voice-overs for nature shows. The male voice-over is manifest in a detached state of interpretive control. Invariably, the narrator speaks as if he were watching with us in the comfort of our living rooms. This tone, too, is an illustration of a state of mind. A detached, remote state of controlled observation. The scripts do offer praise for those animals that somehow survive in this dog-eat-dog world. And they offer us time to think about the mysteries of nature and the cosmic complexity of our world. Unfortunately, we have to complete our thoughts as we skip back into the real substance of TV, that vast majority of programming which finds people pitted against people; that hardy, emotionally involving stuff which supports the weight of commercial advertising.

Plants can't cut it even as filler. They fail to move in an engaging way unless time-lapse photography is employed. Nature shows on TV favour animals because they are more like us. We are animals who watch TV. People are the nature on television. The average North American watches around five hours of people doing things to hurt or help people on television every single day. Other animal species and the backdrop of plants surround people on TV, but generally we are expected to focus on people.

The success of Levine's prediction is not the invention of a viable format of television programming starring the earth's lush foliage.[2] This is not what
he set out to do. Levine was trying to model his state of mind in "Landscape II." As an urban intellectual, a visual artist, he took a kinky approach toward defining a serious problem we all have to confront. How can we know and appreciate nature if we watch television during most of our spare time? How can we experience anything but anxiety when we look at something so alien

as the natural landscape? Most of us live in cities bound tight by architecture. We view the world outside through screens of light. Plant life is boring on television. The idea of nature is intimidating, but a landscape diminished by television makes us anxious for different reasons.

We learn from the news that things are wrong and getting worse. The earth is heating up, the forests are disappearing, the water is full of toxic chemicals—we lose, on average, two species a day. Although we do find our collective state of mind in these descriptions of our deteriorating environment, what do these messages do to help us reintegrate ourselves with nature? In reality, we find ourselves further removed from a direct relationship with the nonhuman aspect of the natural world every single day.

Levine's "Landscape II" frames one aspect of our dilemma succinctly. Most people find the landscape boring on television. He was worried his audience might be too far gone, and rightly so. He faced a difficult problem with a crude set of tools. The cameramen were unskilled and inattentive; the video colour was unnatural; his immediate landscape was impure. But his performance in 1975 was effective in its depiction of anxiety. The information on this recording has remained fresh. His plants would still look beautiful on television today. They will be even more heavenly tomorrow.

> > >

> TITLE AND YEAR OF RECORDING: "Me$$age to China," 1979

> ARTIST: Susan Britton

> BIRTHPLACE AND DATE OF BIRTH: Winnipeg, Manitoba, 1952

> CITIZENSHIP AND CURRENT RESIDENCE: Canadian, New York, New York

> LENGTH OF RECORDING: 21 minutes, 20 seconds

> VIDEO FORMAT AND FORMAL CHARACTERISTICS: 3/4", colour, two-camera "live" recording of performance sequences edited together with pre-

recorded television and Super-8 film, with scripted and improvisational on-camera and voice-over monologue; audio from multiple sources mixed for stereo playback

> PLACE OF RECORDING: Western Front, Vancouver, British Columbia

"Me$$age to China" opens midshot with what appears to be a young Asian woman modelling a traditional "China doll" look. The image is out of focus; the woman's appearance is conveyed primarily through bright colours accentuated by high-contrast lighting. The model is Susan Britton, dressed in a blue Chairman Mao shirt. Her face, although obscured by soft focus, is defined by heavy makeup. Her hair is as black as her skin is white. This impression of a "China doll" is convincing, masking the artist's true identity.

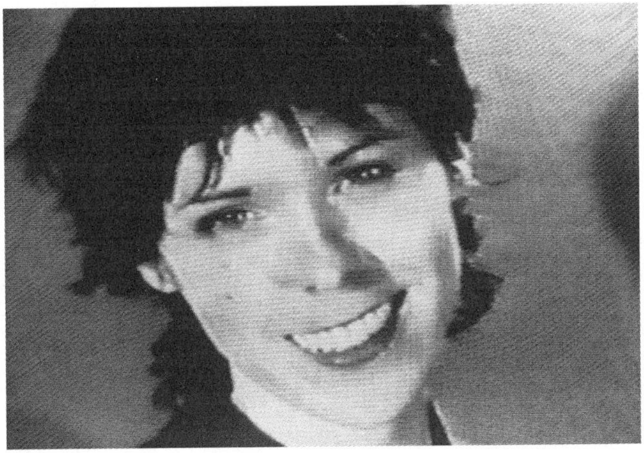

< < <

Susan Britton, as she appeared in her video, "Me$$age to China," 1979, 21 minutes and 20 seconds.

While the camera crawls back and forth and up and down in a slow examination of the soft image of the Chinese woman, we hear the voice of Teng Hsiao-ping, then vice-premier of the People's Republic of China, delivering a speech in Chinese to American president Jimmy Carter and other American officials upon his arrival in Washington on January 28, 1979.

Britton's voice soon joins that of the Chinese leader as she stumbles through a hopelessly inadequate English translation of Teng's speech. Teng Hsiao-ping is now known as Deng Xiaoping, since Pinyin, the system for spelling the sounds of Mandarin Chinese in Roman characters, was introduced later in 1979. Deng (Teng) is now known as the aging Chinese leader who ordered this year's slaughter of students and civilians in Tiananmen Square (Beijing, June 4, 1989).

Britton's purposefully inept simultaneous translation was based on a text published the same week in *Time* magazine. She functionally outlines his message while metaphorically emphasizing the huge gap in understanding between East and West at the time. Britton's groping English translation provides only hints of the meaning of Deng's speech in Chinese: "The significance is mainly ... the normalization of relations between China and the U.S. ... going there ... President Carter and other leaders of course ... be of benefit ... particularly at the present time ... the world is quite untranquil ... where one wants global hegemonism, small hegemonism ..."

The next shot is the title graphic, Me$$age to China, in white characters on a red background. Then a cut back the artist, who now appears as a confident, assertive young Canadian woman standing, looking directly into the camera. She speaks sternly through lips highlighted in bright red lipstick: "What is authority?" Britton walks toward the camera and out of the frame. Cut to the Me$$age to China graphic. Back to Britton, with her black hair and white skin and red lips: "What is privilege?" Cut again to the title and back to Britton; she warms up to the camera as her voice softens: "What is an executive source disk with a zap emulator?" At this moment, an upbeat improvisational light jazz track (piano and vibes) is added to the soundtrack. The video cuts to television images from the steps of the White House in Washington. Deng Xiaoping and his wife, Madame Cho Lin, are greeted by Jimmy Carter and his wife, Rosalin Carter.

Deng and the Chinese delegation are shown arriving for a lavish state dinner.[3] Other important officials, Chinese and American, are shown: Vice-President Mondale, Senator Byrd from West Virginia, Speaker Tip O'Neill, all with their wives, are singled out in the black-tie crowd in a large dining hall. The jazz track is just as bubbly as the champagne as the Americans toast their honoured guests. The image cuts to a video switcher, which is played like a musical keyboard by a woman's hand with fingernails painted red. Cut to Britton modelling with a piece of red fabric. Her voice-over: "The new red banner for now is polyester. Washable, lightweight, and wrinkle-free. We've received your message, you want to learn from the bourgeoisie. Our reply, will that be cash or Chargex?"

Cut to a slow panning close-up of a small soft-cover book framed to look like a generic commodity item on commercial display. The book, standing up on end, surrounded by a platform of ice, could be a plastic container for a perfume or shampoo, or a can of beer, except for its title, which is legible in the frame. It is Joseph Stalin's *Imperialism: The Highest Stage of Capitalism.* The artist's voice-over: "The dialectic moves in mysterious ways. The highest stage of capitalism ... but so is cablevision." Cut to Britton posing like Vanna White (of *Wheel of Fortune* fame) with the video switcher—she caresses the instrument, wearing a "wouldn't you like to have me" smile. Cut to Britton posing with a 45-calibre revolver. "In the seventies, the only alternative was terrorism." She cocks the revolver and hands it to a male model who pulls the trigger of the empty handgun. "Ultimately reactionary and easily absolved ... and you in China were left with a sixty-hour workweek and insufficient hardware. In the eighties, East meets West. Let's re-invent politics together. Hi, China." She waves to the camera. The jazz muzak is perky, supporting her provocative performance.

The title graphic is flashed one more time as a transitional device. The artist's voice over: "Down with nihilism." The image switches to the jazz players: Hank Bull on piano, Eric Metcalfe on vibes (Metcalfe is actually playing a marimba). "In the eighties, East meets West ... down with romanticism." There is a party at the Western Front, the place "Me$$age to China" was made. The happy-hour scene is composed of Kate Craig, Jane Ellison, Glenn Lewis, and Britton. They dance around the jamming players, singing "Hi, China" with their drinks in hand.

We switch abruptly to another space at the Western Front, a makeshift recording studio. A second, different band appears restless between numbers. The message is punk rock, in the late seventies fashion of understatement. The colours are again black and white and red. A single electric guitar and voice. A band of "happy dolls" singing tough songs. Terry Ewasiuk, on guitar, sings about her own forms of self-mutilation. Paul Wong assists with the tech, including some camera work. Britton does a studio interview. She is the leader of the band. She delivers her interview (an answer-only self-interview) in chopped-up sequences. She is drunk on beer or whatever. "We're not going to change the tour. It's really going to go, it's going to make it. We've signed a contract ... we've been in the recording studios for six days." Britton goes on to say she has the hot band that is going to blow China away. They will open in Peking (Beijing). Ewasiuk sings: "Look out you. Look out you, hey, hey, hey."

We snap via straight-cut back into the television footage of Carter welcoming Deng at the White House. Ex-president Richard Nixon was at this dinner party.[4] We hear screams and gunfire in the background. Britton has replaced the light jazz with the action-packed soundtrack of a TV war movie. The combination is chilling, given our recent memories of Tiananmen. Then we cut to some Super-8 film footage of a Chinese New Year parade in Vancouver, with its own muffled sync-sound of crowd noise, drums, and firecrackers. Britton and Ewasiuk are pictured watching from the sidelines in studio shots (video). And then Deng himself takes over the screen. He is big in the frame, in a formal television portrait. Deng's image was taken off American network TV as he sat silently listening to a translator reading one of his speeches in English. Britton's edit has made him appear to be observing Britton and Ewasiuk from his position of unquestionable authority. The "happy dolls" perform for him; Ewasiuk's guitar and Britton's voice: "Remember the chorus, we're just happy dolls, kind of cute and kind of dumb. We're just happy dolls. Pleasure at a maximum ... Okay, we've got a heritage we think you'll like. We'll do anything and never go on strike ..."

Then the "dolls" go silent. The artist adds one more "The dialectic certainly does move in mysterious ways" and "China invaded Vietnam last night" (February 17, 1979). The closing credits are delivered by Britton's voice over the continuing party scene at the Western Front; the cast dancing around the boys finishing up their "Hi, China" composition. The image of Deng's television portrait lingers in memory. He was healthy and strong in 1979, though a bit nervous under the heat of the television lights. America, and Britton, from her production studio on the Pacific Rim, were watching him with wonder.

Britton told the story of the decade leading up to the end of the "pro-democracy" movement at Tiananmen Square *from its very beginning*. When I watched this recording this past summer, for the first time since 1979, I wondered whom Deng would decide to crack down on first. Would it be the "happy dolls" or the regular cast at the Western Front? I imagined a massacre at the Western Front. When, at the conclusion of "Me$$age to China," Britton expressed surprise at China's invasion of Vietnam, she gave this military gesture more importance than it deserved. This is, of course, easy to say today. Hindsight is twenty-twenty. In retrospect, Deng's army was merely responding tit-for-tat to Vietnam's assistance of the Cambodian's attack of the Chinese-backed regime of Pol Pot and the Khmer Rouge. The Chinese felt that the Vietnamese were running all over Asia, and somebody had to teach them a lesson.

Looking back at this event through Britton's tape, the most interesting aspect of China's 1979 invasion of Vietnam, which quickly bogged down and ended seventeen days after it began, was that it was conducted by China's Twenty-Seventh Army. It was this same Twenty-Seventh Army which carried out the massacre at Tiananmen ten years later. When "Me$$age to China" was being made, Britton was focusing on Deng's fear of Soviet hegemonism and his related global strategies. China sought to gain friendship (and economic ties) with the U.S., Japan, and Europe, to neutralize Soviet expansionism. Deng, the revisionist who rose to power following the collapse of Mao's Cultural Revolution, sought strategic allies to assist in the modernization of China. His visit to Washington was part of China's Four Modernizations strategy. The Americans had enthusiastically agreed to lend a hand in the Four Modernizations (agriculture, industry, science and technology, and defence), mainly because such a friendly relationship with China would prove to be worrisome to the Soviet Union. Deng's visit to Washington, the first official visit to the U.S. by a Chinese leader in thirty years, resulted in the signing of bilateral agreements between the U.S. and China covering cultural exchanges, scientific and technical cooperation, a high-energy physics project, student exchange, the establishment of American news bureaus in Peking (Beijing), and a consulate agreement. Although his political rivals within the Communist Party, particularly party chairman Hua Guofeng, warned that the Four Modernizations would not work "without a high degree of political democracy," Deng wasted no time in his drive to modernize China. This modernization continues, despite the setback of the massacre at Tiananmen.

Susan Britton is still interested in politics and media today, though she has abandoned her career as an artist. She now makes her living as a freelance writer in New York. She has left a remarkable body of video work from the seventies behind her. Her video style still has plenty of bite — Britton worked fast and loose and smart. The industrial way she mixed media genres together on shoestring budgets was truly innovative. Her images, framed in incongruity during her literal assault on the medium, are often unforgettable. "Me$$age to China" is part of her legacy — a video history of postmodern thinking charged with insight and irony.

> > >

1 The approach I've taken toward predicting the future by perceiving and describing the present is based on common sense. This formula of prediction is a conservative methodology that assumes if an accurate content analysis is conducted on any current body of information, a probable near future is relatively easy to extrapolate. Campbell, Levine, and Britton all made their predictions based on their analysis of the information available to them in their particular psychological and physical locales at the time.

Campbell's decision to exit Sackville was based rationally on an immense history of artists' gravitation to the urban centre. His formal use of a one-gun video communiqué was a response to the technical limitations of equipment available to him at the time; in this case, equipment available from the Physical Education Department at Mount Allison University. His use of the personal narrative in video was a response to the lack of attention he was receiving in Sackville. He chose to pretend that "the media" was interested in giving his personal reality an audience. The fact that he chose to relocate in Toronto, if only to continue issuing successive statements from a minority position, in various forms of autobiographical fiction, brings us full cycle to the present. Where will Campbell go from here? It is unlikely that he will return to Sackville.

Levine took a different approach toward examining and expanding his base of information. He looked at his immediate surroundings and in doing so clarified his lifelong obsession. Levine found himself living in two worlds simultaneously. It turned out he was not only uncomfortable with the duality of his world; he also proved to himself that things would probably get worse if he didn't reconcile his situation. The landscape tape he made from that television studio in Wayne, New Jersey, must be repeated through other media devices again and again, until it is resolved further. The question is who will continue to work on this problem? Unfortunately, the subject of our relationship with nature is unfashionable in urban art circles today. The problem of our tenuous relationship with nature is increasingly stated through a fixation with the body; the only aspect of nature we can relate to in the city.

Britton seems to have been entirely intuitive when she made her prediction of the present situation in China. But her content analysis of Western media before the making of "Me$$age to China" was astute. She was pulling information from newspapers, magazines, and TV before and during Deng's triumphant visit to America, just as she had previously devoured everything she could find on Mao's Cultural Revolution of the sixties. She had travelled to Vancouver, that Asian Canadian city, immersing herself in China's future first-hand as part of the process of making her tape. In her own words at the time: "East meets West ... Let's re-invent politics together." Deng and Britton have, and they continue to do so.

2 Formal innovation is another area where artists predict the future. Although the limitations of video technology dictate the formal characteristics of all video recordings in practical terms, artists often find previously unexplored potentials in technology generally used for other purposes. Campbell used video equipment normally used to study and improve athletic performance to say goodbye to Sackville. Levine used a television training studio to make a landscape tape, turning a machine normally used for transporting talking heads into a modest version of a space probe (like *Voyager II*). Britton turned an artist-run video production studio normally used for making or documenting art into an international media outlet, from which she released her own statement to China. No doubt if she issued such a statement today, say to any number of countries undergoing democratic revolution in Eastern Europe, she would do so via satellite transmission.

3 The menu for the state dinner at the White House honouring Deng Xiaoping and the Chinese delegation (East meets West through the media of food and drink), according to the *New York Times* (January 29, 1989):

TIMBALE OF SEAFOOD
FLEURONS
ROAST STUFFED LOIN OF VEAL
SAFFRON RICE
BROCCOLI SPEARS
ENDIVE AND WATERCRESS SALAD
TRAPPIST CHEESE

CHESTNUT MOUSSE
CHOCOLATE TRUFFLES

PAUL MASSON PINOT CHARDONNAY
SINU ROSE OF CABERNET SAUVIGNON
HANNS KORNELL EXTRA DRY
CHAMPAGNE

4 Again according to the *New York Times,* this visit was ex-president Nixon's first to the White House since his tearful farewell on August 9, 1974. Henry Kissinger also attended, though he wasn't listed on the official White House guest list.

Thanks to Alan MacKay, then director of Art Metropole, for commissioning this research; and Susan Watterson, MacKay's successor, who followed up on this initiative; and to Amy Maggiacomo and Michael Klein, then staff of AM for providing logistical support during my research.

The Face of the Machine 2000

Granular Synthesis, the name adopted by artists Kurt Hentschläger and Ulf Langheinrich, is derived from granular sound synthesis, an information-processing technique for synthesizing digital audio. A series of short samples (grains) are sequenced, reassembled to produce a granulated sound synthesis. Selected whole sounds (and/or images) are fragmented into tiny snippets (grains) and are recombined to make new granulated sound or image continuums.

Hentschläger and Langheinrich apply and extend the primary technique of granular synthesis to transform the gestures and voices of human bodies, particularly human heads isolated by video from their bodies, into recombinant, alien screen-based creatures that move and express themselves through suprahuman performance. By breaking down video recordings of human models into split-second frames of video and sync-sound audio, and then reconstructing these models from these fragments, making them whole again (but giving them a new range of behaviour that dislocates them from their pregranular nature), we witness humans evolving into machines. In short, Granular Synthesis transforms people into machines.

We already know that we are becoming machines, for a cyborgian self-image is now available nearly everywhere. Cell-phone antennae protrude from our heads. Computer screens pop out from our laps. Experiencing the work of Granular Synthesis is like watching the accelerated evolution of our species through time-lapse photography. It doesn't take long to see where we are headed. There is real beauty in the horror, when that which is left behind — traditional humanness — is juxtaposed with the power of digital electronic gain.

Granular Synthesis first showed interest in fragmenting and recomposing moving pictures and sync sound in a 1991 group exhibition, at Kampnagel KX, in Hamburg, Germany. Hentschläger and Langheinrich appropriated and remixed an underwater scene from the B-movie *Piranha* (1978), shuffling the shots and looping the remix. The cinematic logic of the scene (a victim is eaten by a school of piranha in extreme close-up) was reconfigured into a state of pure, suspended chaos and anxiety.

In 1992–93, Granular Synthesis began a series of performance works, each progressively refined, like a series of computer software updates: Modell 3, 3.01, 3.02, Modell 4, and so on. As this work progressed, it became clear that the video and audio arrays of Granular Synthesis are a kind of software for programming an audience, controlling the perception, experience, and ultimately the descriptive processes (conscious reflection) of the audience. Granular Synthesis creates precise scripts, new forms of strict time design, for its audiences to follow and internalize.

The bodies, or more correctly the body parts, on the video screens are fragmented, then reconstructed in clusters of fragments, like "time cells" of human appearance and performance, some no longer than a few video frames (1/13th of a second), but none much longer. We can normally only consciously differentiate between four to five frames of visual information per second, so much of the texture of this granulation is read subliminally. The ear does much better in its conscious awareness of the texture of fragmentation and in determining the relative, perceptual bump and grind of passing "time cells." The viewer's ears often tell the viewer what his or her eyes are seeing.

Granular Synthesis's performances isolate and dislocate appearance and behaviour from the cause and effect logic of cinema, television, video, music, and our perceptual experience of "full-motion" life itself. Time-based technological media, film-video-music, are generally programmed to run parallel to and analogously alongside "full-motion" physical reality. Although there is an extensive history of time alteration in film-video-music, the time-based arts—we've learned to read flashbacks, flashforwards, the instant replay and more and more variations of time-event compression and density—Granular Synthesis's machine goes further and is designed to totally shatter the full-motion status quo.

Throughout the 1990s, this creative research progressed through "Modell 5" and finally into "NoiseGate-M6," an incredibly intense and beautiful installation released in 1998. "NoiseGate" is an installation, not a performance work. As a comprehensive, totally immersive information environment, it offers an experience fixed in a specific, concrete space and is open-ended in duration. Granular Synthesis has refined "NoiseGate" in multiple installations and versions over the past three years, while concurrently performing "POL," the latest performance work. Both "NoiseGate" and "POL" are more abstract, less literally figurative works than the work from the early and mid-1990s. Increasingly, Hentschläger and Langheinrich have shifted the focus of their software from the manipulation of human models represented in their video to more directly targeting, pushing, and perceptually challenging the audiences of their performances and installations.

With "NoiseGate" the Granular Synthesis machine is once again applied to the task of shattering the apparent reality of a human subject, but this version of the machine is vastly more powerful and the framing, confinement, and isolation of this human subject is also far more precise and definitive due to the introduction of a new architecture of containment or imprisonment. A new factor of control, the idea of the cage, has been added to the formula for measuring and studying the parameters of human behaviour. We now witness an even greater isolation, magnification, and amplification of the untimed, denaturalized, or denatured human model.

The initial analogy, encaged zoo life, pacified and specially illuminated for the guilt-free pleasure of our observation is replaced by a single human exhibiting a range of behaviour in multiple phases. The dark recesses of this unusual human zoo form a thick ambient noisescape, which rumbles and recedes into the visitors' subconscious, remaining only as a sea of sonic pressure as the heads-faces speak with their drifting, deliberate behaviour sampled and reassembled with the characteristics of the system.

With "NoiseGate," we have entered a more abstract zone. The machine has generated a field of data composed of multiplied, magnified, modulated, memorized human behaviour. This field attracts and collects variously associated physical, psychological, and social phenomena. This new machine display is an association generator: an instrument for the imagination.

"NoiseGate" invites us to navigate an environment where humanness is determined precisely, magnificently by a machine. We wonder how to interpret the data without the aid of additional instruments.

The encaged human behaviour drifts in and out of phase, pulsating and cresting in waves of resonance in a system oscillating with a life of its own. People enjoy the experience of such elementary polyrhythmic movements when these movements are not forced on them—when the surface of the ocean moves or when we see an animal's rib cage panting, or when the rhythm of sex fuses two bodies into one, or when the mantra comes together to invoke a sense of peace and well-being.

The life rhythms of "NoiseGate" are artificial and imposed and generate feelings of dislocation and neglect. Granular Synthesis has created an environment where the individual is left entirely to his or her own devices, abandoned in a state that is totally controlled. How would one survive in such a state? Lashing out in direct, immediate reactions to intolerable aggravation cannot be sustained. Eventually, there will be disinterest and indifference—clear signs of resignation. More chronically, when the subject of such a state realizes that violent outbursts or even plaintive cries for attention have no effect, there will be a progressive withdrawal and decline and

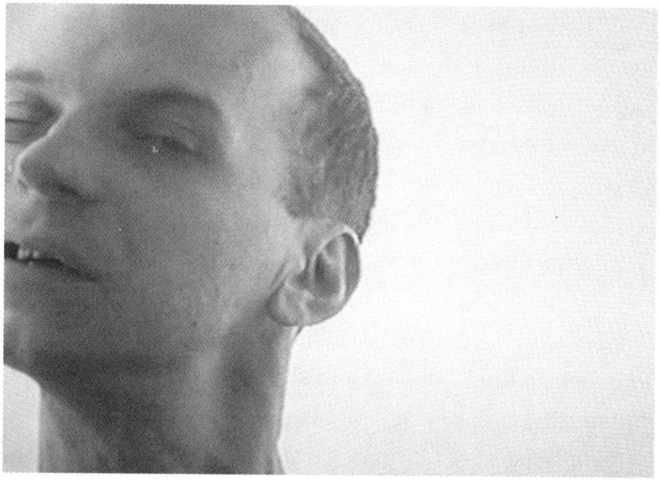

> > >

Michael Krammer, as he appeared in "NoiseGate-M6," a video installation by Granular Synthesis, 1998. Photo: Kurt Hentschläger eventually total passivity. From restlessness to passivity, with occasional outbursts or spasms of aggression, eventually directed against oneself, the only object left.

"NoiseGate" then becomes a timeless space for reflection, for associations made against a backdrop of an unrecognizable drone. The human face of the machine looks out at the visitors, and they see themselves in its blank expression. The vacant face holds tight. The face becomes the insistent logo of the disappearing human being.

We are left with an internal dialogue between organic-analogic and artificial-digital systems linked and mediated by our human senses. We stand nakedly human as an organic perceptual system confronted by a challenging posthuman environment. Standing alone, immersed in "NoiseGate," we "read" the literature of Hentschläger and Langheinrich to the best of our ability, a literature written by a machine with a human face, a literature composed of immensely complicated terrains of remote-controlled behaviour. With the gain, the amplification and expansion of human nature, there is loss. There is an increase in noise, a loss in the central clarity of being human.

This distortion should not be viewed as a shortcoming of the system, for nothing essential ever happens in the absence of noise. Viewing noise as degenerative disorder is not useful when considering Granular Synthesis's work. There is no absolute structural difference between signals and noise in their work. The recombination of bits of reality without ever imposing a structural overview is merely an admission of an essential paradox. Whether we cast "NoiseGate" in the light of our evolution from humans into machines, or look carefully at how waves build, crest, fall apart, only to immediately begin forming again, there is one primary set of recurring ideas. In all gain there is loss ... and in degeneration there is regeneration.

ON THE FUTURE OF ART

TOM SHERMAN

Tweak 1995

Tweak. Tweak is the title of this thing. To tweak one's perception or experience or description of life is art. Art is communication. It is not communication. Art is information. It is not information. Art is stereotypical; it is atypical; it is prototypical. Art is speculation on the future; it is speculation on the past. Art is an accurate description of the present. It is a language for defining oneself. It is a language for finding others like oneself. Art is a language for immersing oneself into a global society. Art is useful for distinguishing oneself from everyone else. It is a language for communicating with other species, a language for distinguishing us from machines. Art is a substitution for one's presence in the world, a house one builds to provide shelter and a sense of identity. Art is the way people reproduce without producing dependants. Art is a way of escaping reality, other people's reality. It is a refusal to acknowledge other people's expectations. Art rejects the feeling of belonging. Art is an embrace. Art is the quest for immortality. Art is an expression of mortality. Art is an admission of need, a declaration of desire, and sometimes a statement of satisfaction. Art is a distraction, it is a lie, it distracts others from our private self. The artist must keep his or her private voice private. That is why it is better not to make art, if one needs privacy. One may find more privacy in silence.

Art is the reconstruction of self. Art is the transformation of self into community. Art is the devolution of self into personal substance without consciousness. Art is the transfer of consciousness into inanimate objects. Art is the description of experience, the manifestation of perception. It is an aggressive act of violence against others. Art is an attempt to bond with every other person on the planet. Art is life, the creation of life. Art is death, a celebration of letting go. Art is everything I am not. Art is the creation of life beyond the body. It is the creation of life, a life that can sustain itself, on the attention of others. Art can be used to defend oneself or to harm others. Art is an accident ready to happen. Art is the way dead people communicate with the living. Art is the way the living speak to the dead, the way people without friends find comfort and security, the way people escape from a suffocating stew of global identity. Art is the way people escape their skins, real and imaginary. It is the way humans fly, or live underwater, or in outer space. It is a way to walk through walls or travel faster than a speeding bullet. Art is a way to become tiny, even microscopic, or huge. Art is a way to become gross and obnoxious. It is a way to become perfect. Art is a way to transform the self into another form completely unrelated to the body. It is a way to become invisible. It is a way to be in two places at the same time. Art is a way to become a machine. Art is the way to become a planet instead of a person. It is the way to become a galaxy. Art is an endangered species of language which must be saved. Art is extinct. The artist is a conservationist, attempting to save art from extinction. The artist is a greedy hunter, trying to kill men without art, whomever they are, or wherever they may be. Artists make art that they themselves collect. They collect themselves in their own galleries and museums, in their studios and homes. Making art is a form of conversation with people who are nowhere to be found. Making art is an excuse for not spending time with other people. Making art results from feeding on other people's art. If art isn't needed today, maybe we are wasting our time making art? Even if art isn't needed today, it is still very important to make art so that when people need art in the future, there will be art for the needy. If art isn't needed today, the art that is being made today must be being made for another time. We seem to be making art and storing it so we can have enough for a period of time in the future. The audience doesn't need art as much as the artist needs to make art. Artists need to have art around them.

Art seems to be getting smaller and lighter these days. It often seems to be completed in the planning stages. Art is so compact today that it takes secret codes to unpack it, restoring it to its full size and impact. Art is a mystery, it draws you to the scene of a crime, and then it demands that you find out who done it. Art is a documentary. It tells us a story about something. Art is impressive because of the research behind it and the organization of information that determines the way the story is told. Art is a kit for making art. Those who stumble onto art learn how to make art from the art they find. Some art, the most impressive art, is art that reproduces itself, art that makes art. Art is a machine, like a computer, which if given the proper instructions, can take data input from everyone, everywhere, anytime, and it can somehow transform this data stream into information. It can make sense out of nonsense. Art is a set of instructions for making art. It is a script for living. Art is a self-authored obituary to be inscribed on the artist's gravestone. Art is the remembrance of something we used to need. Art is no different today than it was in the past. Everything else has changed, but art has stayed the same. Art is an anchor to hold us down against the winds of change. Art is a signal, sometimes a very weak signal, of things to come, things we don't recognize yet, because the time has not yet arrived. The artist is an advance scout, who rides out ahead, blazing a trail for us to follow. The artist is the antennae of his or her race. Art is speculation on what is happening today. It is speculation on what may happen in the future. It is speculation on what has already occurred but has been forgotten. Art is all of these things simultaneously. Art and artists are mixed up and have no idea what they are doing, or where they are going or even what they have accomplished. Art is the only natural communication the species undertakes. Art is the most unnatural communication the species attempts. Art is a code that permits secret messages to be exchanged without being intercepted. Art is an environment that artists create to live in. It is extra space that artists need, without their art they feel hemmed in. Art is a virtual space, not a real space, that artists produce to exist in. Art is a trap door, which opens up into a different dimension or reality. Art is dead, over, done with. It is obsolete, like a technology or a species that has gone extinct. Art is an escape hatch or a pressure valve; it lets the artist blow off steam. Art is a series of new toys; artists make things to amuse themselves. Art is the manifestation of guilt, or greed, or fear, or generosity. Art is the hard copy of love. Ultimately, it is all that is good, a generosity of spirit.

Art has been various things to people at various times. As the times and the needs of people have changed, art has changed, and thus the role of the artist has been redefined. In this era, which is marked by extreme technological change, radical change, artists are being forced to be more conservative, to function as a force for societal stabilization. There is one thing we can say for sure, art is something that can be used to influence people. Politically correct art is made and used to influence people to become more politically correct themselves. Art is made to advance political change. This fact has led to socialist art in capitalist countries. This fact has led to art expressing individualism, perceptualism, and formal concerns in socialist countries. In North America, we see much more interest in celebrating diversity and in recognizing the other: women, underrepresented groups, native peoples. This trend may continue with the recognition of handicapped, or physically and mentally disadvantaged individuals, or the recognition of the underprivileged, the uneducated, the creative spirit of the community, the neighbourhoods, the folk artists, the naive. The North must recognize the South. Concern for the environment and respect for the diversity of the biosphere will prevail in all aspects of culture.

The idea of the naturalist, the artist as naturalist, the one who makes things about what he sees, honestly, without trying to shape his reporting to favour only the beautiful or the grotesque, will emerge. The idea of the naturalist is indistinguishable from the perceptualist, the one who represents his perceptual experience. The impossibility of feeling that we are one with nature is a good problem for the contemporary artist to grapple with. An artist will be less concerned with the depth of his or her work than with its presence in the media. Art should function like a sound byte. It should find its complexity and scale of influence in the presence it has in the media. Instead, we see artists amassing huge collections of their own work, often repeating the same gesture over and over so it may accumulate in a pile of physical objects, a series of unique physical objects, with a tiny percentage of these things reproduced in media, only very rarely, if at all. Tweak, tweak, tweak, tweak, tweak, tweak.

TOM SHERMAN

Thoughts from the Antechamber (On the Future of Art) 1995–96

Imagine we are in a room, a waiting room, an antechamber. We are waiting to enter the next room, the big room, a huge room that is dark and silent and unknown. This foreboding room is the future. In 1995, we find ourselves together in the antechamber, waiting for the close of the twentieth century, waiting to commence the next millennium. What does the future hold? How can we best prepare for reality in the next room?

Some of the people in the antechamber are artists. Actually, there are quite a few artists. They begin to talk. What will the next room be like? They agree on certain things. At the turn of the century, the world will continue to be ruled by men without art. There will be more and more technology. Our relationships with people will be mediated to the greatest extent possible by technology. Human contact will be less and less direct. Our species will progressively have a more and more difficult time staying connected with nature. We will worry a great deal over losing touch with nature, but nature itself will shed no tears over our dislocation from the natural world.

We are leaving our bodies behind. It excites us when our blood is transformed into electricity.[1] We love speed — to be as fast as light is our ultimate goal. To get there we must abandon our bodies. We must become the energy and light driving our most impressive machines. We will be magnificent in our new immaterial form; in perfect accord with speed itself. We will be lightning-fast. We will run super cool. Only nature will remain unimpressed. Nature's significance will fade — it will become a mere backdrop for our flashes of brilliant illumination. Our blood will be transformed into electricity. The light of this transformation will be blue. In the colourless vacuum of the big room, the dark silence of the future, we will all be blue. Despite our many differences today, tomorrow we will all have blue blood. Our machines will treat us equally as human beings. Once we are blue, we will be quick and cool, free; disconnected from nature, altogether.

Getting Connected, Staying Connected[2]

So we begin to talk to each other in the antechamber. We put down our magazines and newspapers, and we stop watching the news, weather, and sports on television. Industrial media formats certainly have their limitations. We talk instead. One strategy emerges from our conversations. We have to get connected. We will have to get connected, and we must find ways of staying connected. Our conversations must continue when we hit the big room. When we cross the threshold into the expansive unknown of the next century, we will need to be connected with others. Getting connected, staying connected; that is our first strategy.

Men without Art Rule the World

The big problem is men without art rule the world and will continue to do so in the foreseeable future. The greatest number of people are disinterested, if not hostile, toward artists and art. Given this reality, communication between artists or between men and women with art is important: marginal, but important. We must continue to build and maintain our subcultures. Identifying exclusively with others in the art world has been common behaviour in the twentieth century. There have been few opportunities for creative work outside the art world, so developing concentrated communities in our marginalized art subcultures made sense. In the next century, things will get much more challenging and hopefully more interesting. There will be more opportunities for artists outside the art world. In the twenty-first century, creativity will be viewed as a valuable resource, like oil or fresh water. But creativity is one thing; art is another. Creativity as a commodity has mobility. Art is more concrete and bound to specific territories. Art will have to be redefined to be meaningful and effective outside the territories held historically by the art world.

Art will be perceived as information and knowledge in the information economy that will define the twenty-first century. Art will be commodified as information or as a set of behavioural roles for the leisure society. This is the next pigeonhole.[3] The leisure class will include youth, the elderly, and all those others branded unemployable by managerial science. Managerial science is scientific management based on information and knowledge.[4] Knowledge in this context is defined narrowly as know-how or how-to (the ability to apply information to solve problems and/or to provide information-knowledge services). Productive work will be strictly associated with managerial science. Art will be a kind of surrogate or default "employment" (more accurately described as societal engagement) for those without a direct role to play in managerial science.

The Split between Managerial Science and Art

Managerial science has captured virtually everything that is not art. In the information age, every human pursuit is approached systematically, scientifically. Sports, fast foods, dating services, higher education, traffic management, politics, law, welfare—virtually everything is governed by managerial science these days. In a pan-capitalist universe, every imaginable activity is governed by scientific management based on information and knowledge. Intelligent decisions are made by informed decision makers. To this end, huge volumes of data are crunched with more power every day. Computers are faster and cheaper by the month. By the year 2000, computers will outnumber people on the planet. Computers and scientific management go hand in hand. Science used to be pure and abstract and detached from the world at large. It existed in the descriptive languages of mathematics, physics, chemistry, biology, and astronomy. Science was once experimental, intuitive, imaginative, philosophical, and theoretical. Now, science is everything that is governed by scientific management (it too is managerial science). Pure science is history. Pan-capitalism has absorbed science just as it has absorbed everything else. Science today is a global community of information and knowledge workers, a subordinate workforce of the global state of managerial science. Facts and figures, statistical analyses, and projected outcomes are the heart of the new, all-pervasive world of managerial science. Managerial science has become virtually everything that is not art. The split is no longer just between art and science. The split is now between art and every other kind of work that people do.

Strategies for Reintegration (Escaping Marginalization and Exploitation)

If artists and art cannot function effectively in a world dominated by managerial science, what can be done to make things whole again? This question has to be asked, even though it is a privilege to exist in a zone at least temporarily excluded by pan-capitalism. How effective can artists be if they reside exclusively in an economically dysfunctional sector of society? To answer these questions, let us begin by asking why haven't artists and art been absorbed by managerial science like everything else? Artists and art are difficult to make more manageable, systematic, and orderly because artists are generally not interested in adopting common goals or in conforming their behaviour to meet corporate objectives. The qualities of art always remain relative, not fixed or explicit. Ambiguity is a virtue. The meaning of art is rarely concrete. As charming as nonconformity, ambiguity, and volatility may be to some, these playful characteristics confound and frustrate managerial scientists. The reintegration of artists and art into a world governed primarily by economic concerns will have to be accomplished through economic integration. Maintaining economic disintegration and dysfunctionalism through anachronistic creative behaviour will be easy. Art's pigeonhole has been defined by history. The economic reintegration necessary for art's expansion into new territories is a vexing problem.

In an era dominated by an emerging, maturing information economy, how can the work of artists be seen to have value as information and knowledge? Art is not widely perceived as information in the late twentieth century. In an economy based on the production and exchange of information, art must be seen as information if the artist is to be reintegrated into society. For that matter, pure, creative scientists (experimenters and theorists) face a similar fate of exclusion. Information and knowledge are defined so narrowly in a world governed by managerial science that it is difficult for people to make connections beyond the connections that computers can make. The all-pervasive mindset of managerial science gives the edge to machines. Works of art generally do not resemble or function as machines. Artists generally do not act like machines.

If artists are eventually categorized as information and knowledge workers, as professionals who make and apply information to decision-making processes valuable to themselves and others, they will then be integrated into the information economy and into a society governed by managerial science. There are probable ways this will happen. There are obvious shortcomings in our societies; shortcomings that will have to be addressed with appropriate information and knowledge. There are serious identity problems for all people living in a world governed by managerial science—a world dominated by machines. Automation makes us more efficient, but it does not help people acquire self-knowledge. Artists, on the other hand, provide a range of potential human behaviour through their lifestyles, behaviour, and art. Artists offer multiple models of identity, hybrid manifestations of human nature, reminding others that human beings, not their managerial systems, are the authors of the human condition.

Identity is natural, physical, personal, individual, corporate, local, regional, national, global, and cosmic. Identity is a complex self-awareness made up of shifting, multiple perspectives of self. Identity is a dynamic phenomenon—it is constantly redefined by changing physical and social contexts, biological development, aging, and perceptions of status. Managerial science produces inflexible, industrial models of identity largely through commercial entertainment. Social, political, and thus economic control are the goals of all industrial identity campaigns. A consciousness industry without artists and art is an industry destined for failure. For our societies, this failure is marked by human tragedy. Mental illness and violence are the chief byproducts of the industrial consciousness industry. Artists offer multiple models of identity, a greater range of possibilities for human growth, and ultimately freedom from the confines of conformity characteristic of industrial models of identity.

Identity is developed, cultivated, and transformed through experience—shared experience. Groups of people communicate over common interests and circumstances to form communities. Communities formed around art, the making of and serious consideration of art, are called art communities. Art communities are loosely woven together to form the art world. The art world still exists separate from pan-capitalism's primary, transformative instrument of order, managerial science. The art world is by nature a highly romantic global society. If fact and figure are the measure of managerial science, imagination and emotion govern the art world. The art world is a coalition of impractical, disenfranchised individuals, many of whom perceive themselves as victims.

The art world reaches out to others, not surprisingly. For its members have strong beliefs and ideals, if little economic clout. They consist of those who have made the choice and have rejected managerial science and all others left out for one reason or another. And there are so many others. The art world actively courts others wherever the others reside. Pan-capitalism has its base of operation in the northern hemisphere. To the south lie the great masses of "underdeveloped" peoples: billions basically ungoverned by pan-capitalism. Add to these massive populations all the remaining aboriginals in remote spots here and there. Plus, back in the urban north there are the racial and ethnic minorities and the gays ... and there are the women and children everywhere ... and the elderly. All these others are embraced by the art world because they too are excluded and exploited destructively by managerial science.

The art world is made up of those who have decided to reject or have been rejected by managerial science, and these outsiders reach out to the innocent and naive. The setting for this romantic community is the garden: nature itself. The art world inhabits the wildernesses of the world's cities, the vacant lots where the wildflowers grow, and it reaches out to the mountains, forests, plains, deserts, and oceans of the world. This is why all the animals and plants on the planet are potentially part of the art world's legions. Artists are like children dreaming. The wind, the rain, the sun, the moon, and the stars still belong to today's children. From birth to the tenth year of life all children are directly connected to the forests and jungles, and their imaginations run wild and their emotions flow freely. In the tenth year, mostly young men and women without art emerge, abandoning the forests and jungles, feelings and dreams, for practical reasons. Most men and women give up their childhood in their tenth year, becoming preoccupied with systematically working their way into adulthood. The vast majority manage to work their way into a world without art. As they claw their way out of the underbrush of exquisite uncertainty, they celebrate their rite of passage with banal declarations of loss. Their chorus is so dry, a perfectly modest refrain: they can't draw a straight line without a ruler; singing is for the birds; dancing is something you do in a bar before you pass out drunk. They are afraid to play games without rules.

The artists join hands with the homeless, the aboriginals, the masses of the third world, people of colour, the gays, women and children, the elderly, and the endangered species of the world. They circle and embrace a planet piled high with human waste. This love of the land and the vast populations of the excluded is an attempt to touch something real, to connect with others who are not the enemy. Men without art do not remember the days before progress and increased productivity. The enemy cannot recall the days of their childhood: days when art was not a thing for the talented few, but it was natural for all to enjoy making pictures and singing at the top of their lungs and swinging their arms out wide as they spun themselves silly in dizzy circles, falling down laughing in a contagious rejection of proper behaviour. That was being a kid and being creative, naturally. Or remember the pain of undergoing a rude awakening, the shock of seeing people suffer, not getting what they deserve or want or need. Remember encountering injustice and expressing disgust. There is a vast experience outside the world of the efficient and effective and the oppressive vulgarity of conformity and complicity.

All this sensitive crap is just that in a world governed by men without art. It is entertainment for the leisure class: there is plenty of room for creativity in a massive inventory of escapist products or in software for engineering industrial-strength individualism. Despite what some think, commercial television, movies, and popular, industrial culture are not the art of today. Art is surely not a redundant code for brainwashing people into the mindless consumption of material goods or an instrument for engineering a massive conformity of fantasy. Art has never been industrial in scope. There is no generic art, an art for every man and woman. People are not so common. The industrial consciousness industry rejects the notion that art exists in a myriad of forms. Art with a capital A is stereotypical Art. Art with a capital A is defined by history; Art history. And even stereotypical Art is not attractive to the broadest public.

Most stereotypical Art is based on the romantic idea of individual genius. Art is made by special, unique people who do not fit in the world of managerial science, or in any world for that matter. These special people, too, are outsiders: either ahead of their time (if already dead) or far too sensitive or psychologically unstable to function in the world of managerial science. Such obvious, stereotypical Art can be and often is dismissed simply reactionary individualism (a modest resistance by a tiny minority of special others). This kind of Art is just as easy to embrace as it is to dismiss. This crazy, one-of-a-kind Art is easy to pigeonhole, either as wonderful, enriching anomalous behaviour or as a threat to the preferred, conventional values of men without art. If necessary, this kind of Art will be easy to take out. Such obvious Art is not even a moving target. It is fixed as a marginal behaviour in a studio loft somewhere private, or in a concert hall where the audience is elderly and sleeping through the spirited performance of their favourite compositions. Such Art with a capital A is made by people wearing leotards jumping around in historic properties.

This stereotypical Art is obviously not alive, though it is still better than nothing for the Art lover, which is what we may end up with tomorrow: we will very likely end up with nothing called art. Artists are an endangered species. Art is being pigeonholed as the something confined to the domain of the other. Professional artists are being forced to exist in a zone surrounded by pan-capitalism and managerial science. As tragic as this outcome may seem, is this a good enough reason to be inclusive to such a degree? Appearing so desperate is certainly a sign of weakness! Remember, artists are trying to align themselves with all the world's others, including the kingdom of animals and plants. The odds are not good for the others, nor for artists. As managerial science advances technologically, artists are destined to work with the primitive technologies of the third world; they must embrace the traditions and histories of vanishing civilizations.

If there is so much hostility or indifference toward art, and the vast majority of artists "choose" to live in the past, then industrial culture will soon completely replace art as a commodity or service for the leisure class. Industrial culture governed by managerial science will all but replace art. If art is to survive in the world of men without art, it will be forced to be safe and predictable, basically resembling in tone the escapist territory of art's overwhelming competition, popular commercial culture. Art that seeks the audience of others, an art not totally superficial, but deep, will be ghettoized; it will be relegated and confined even more deeply in the pigeonhole of art. In the economic framework of the era of information and knowledge governing the twenty-first century, art will be a marginal subsector of entertainment. (It will be defined as difficult, sometimes unpleasant "entertainment.") Art will be the principal device for representing and interpreting the other: all those excluded from a rapidly advancing world governed by managerial science.

Nothing Called Art

So the artists have gotten into an abstract conversation about what art may be in the next millennium. The antechamber is a serious place. It has to be. There is so much uncertainty and danger. The problems are hard to grasp and even more difficult to face up to. Some artists in the antechamber have suggested that art may disappear altogether; that art as we know it today is probably headed for extinction.[5] It may survive simply as a nonretinal phenomenon, identifiable as art for a while longer, another seventy-five to a hundred years. It may continue to be nothing more than ideas or concepts, as it has been for the better part of this century. Underneath art's colours and precious materials, art has been and continues to be fundamentally transparent, lurking as a code in every territory but that which is too obviously art itself. Art, in this way, has been invisible for a long time. So few have noticed its "disappearance," preferring to believe that nothing has changed; preferring to believe we are living in the past.

Most look for stability in art and take comfort in art's conservative nature; its stereotypical appearance. Art history has become art (the inversion of art becoming art history). Art has become a manifestation of memory and tradition for most. For others, art has become a theoretical exercise, a declaration of ethereal speculations. Art will undoubtedly continue to be a projection of theoretical concerns, an analysis of the basic parts of something incomprehensibly complex: a single description or two, or a suite of descriptions of something that cannot be seen whole or will likely never be obvious in its full complexity. With theory-based art, artists merely scratch the surface (of a new body of knowledge). Artists ask questions that cannot yet be answered. Or put another way, they offer answers that must be questioned. In an information economy, a world articulated by information and knowledge, good questions have value. (Some managerial scientists may actually relate to such art!) Questions cannot be answered (responded to) unless they are posed.

In the immediate future, artists will author navigation systems, and their art will function as probes for organizing or cutting through vast fields of data and information: incomprehensibly complex fields or, perhaps more vividly, massive walls of noise. In a universe defined by relationships between people and machines and ultimately machines and machines, self-identity must survive a masking and mutilation of the senses. The species will depend on a handful of individuals with the strength to hold onto the raw information at their fingertips, no matter how hostile the environment.

Managerial science is describing everything with facts and figures. Everything is mediated in degrees (levels), and direction and proximity bear little relationship to amplitude and frequency. Nothing adds up unless you are heavily automated. So many people are flying blind, completely dependent on instruments. Look out. The machines are networked in gigantic webs, forcing collective insensitivity, among many other things. What a buzz. Numbness is confused with experience. We need a few clear-headed souls to guide us through the thick fog of disorienting signals.

In the antechamber, there is only memory and hope. Life, what's left of it, is simple. There are anarchists and archivists. Yes! There are still artists of different stripes. Some will recombine their own histories until their self-identities are as complex as a city full of bright lights on a warm night. Others will clear away the self as they describe something alien and inhuman; something "natural" in the inanimate sphere, like an idea or a concept. Artists and their art will migrate out of their physical bodies and objects into a new world of atypical immediacy. The rare executive of managerial science will welcome these contemporary artists and their fresh, new provocative art with open arms. Information and knowledge cannot exist without questions that generate new responses and breed further questions. Artists and art scratch the surface revealing codes transparent and ubiquitous. Artists will ask the questions that children ask: the questions left unanswered out of kindness. Art will disappear for all but the most curious. The men without art will be left with nothing to ignore or complain about.

1 Arthur Kroker and Michael A. Weinstein, *DATA TRASH: The Theory of the Virtual Class* (New York: St. Martin's Press, 1994).

2 "Getting Connected, Staying Connected" was the title of a conference that never happened at Syracuse University. GCSC was the brainchild of Jim Leonard, a student of SU's School of Art & Design. Although the conference was cancelled because of financial shortcomings, many of the ideas generated in preparation for GCSC were realized in publications, exhibitions, and seminars at SU in 1995–96.

3 The emergence of the "tourism and leisure" sector of the information economy is normally understood as an information-entertainment sector for those

privileged enough to have the time and income necessary to travel and play. The strange paradox in this equation is the confinement of others in the leisure time-zone (teenagers in shopping malls and video parlours; the elderly watching television forever or escaping on banal bus-tours, and, of course, the fact that non-Western cultures have massive amounts of free time in their "anachronistic" cultures). For a good overview of the definition of a leisure sector of the information economy, see Frank Feather, *G-Forces: Reinventing the World, the 35 Global Forces Restructuring Our Future* (Toronto: Summerhill Press, 1989).

4 Herbert Simon, *The Sciences of the Artificial*, 2nd ed. (Cambridge, Mass.: MIT Press, 1982), 34 (under Procedural Rationality): "Today several new branches of applied science assist the firm to achieve procedural rationality. One of them is operations research (alias management science); another is organizational theory. Operations research provides algorithms for handling difficult multivariate decision problems, sometimes involving uncertainty. The simplex method for linear programming is an important algorithm of this kind; queuing theory is another; linear decision rules for inventory control and production smoothing are a third. These and similar computational procedures are widely used in business decision making today." For a brief survey of these developments, see H.A. Simon, "On How to Decide What to Do," *Bell Journal of Economics*, 9 (1978): 494–507. For an estimate of their impact on management, see H.A. Simon, *The New Science of Management Decision*, revised edition (Englewood Cliffs, N.J.: Prentice-Hall, 1977), chapters 2 and 4.

5 Art will appear to become more and more like itself in the twenty-first century. The traditions and conventions imposed by art history will pigeonhole artists into a realm of limited behaviour. The art pigeonhole is already surrounded by managerial science, pan-capitalism's ultimate instrument of control. Artists may choose to exist in this anachronistic art conservation area as an endangered species, or they may evolve strategies for escaping the pigeonhole. Perceptual and conceptual strategies, the underlying code of all advanced art in the twentieth century, offer multisensory and nonretinal paths into underdeveloped art territories in the twenty-first century and beyond.

The multisensory path: digital multimedia has potential as a return to perceptualism, as a multisensory, immediate, interactive rejection of the conceptual and theoretical path. Ultimately, digital multimedia is going to run parallel to analogical perceptualism (postliterate or postconceptual awareness). Virtual reality, as the potential zenith of digital multimedia, will amount to the detailed construction of a realistic second nature.

In the nonretinal approach, conceptual paths are tied to fading print literacy and theory, perhaps a domain of an elite holding onto an abstract, arcane concept of privacy with accompanying sensory deprivation. Nonretinal art, whether alphanumeric, oral, or acoustic, is a denial of visual sensuality. As such, it offers a clear option to the self-imposed limitations of visual art, that art looking more and more like art every day in the art pigeonhole.

Vision Beneath Light 1997–98

In 1995, I began writing a text called "Thoughts from the Antechamber (On the Future of Art)." It turned out to be a dark text describing the marginalization of art by managerial science at the end of the twentieth century. As a retrospective overview at century's end, it was designed to open a window, an escape hatch, permitting artists to evade the pigeonhole of art (the art of historical reference — art that is unquestionably art in a conventional, traditional, historical sense). For clarity, let me repeat and expand these two strategies, touching on their limitations and potential:

Option 1 — The Nonretinal Approach: Nonretinal art, whether alphanumeric, oral, or acoustic, is a denial of visual sensuality, and as such it offers a clear option to the self-imposed limitations and vulnerabilities of visual art, that art looking more and more like art every day in the art pigeonhole.

This option is perhaps the shortsighted approach, as younger audiences consistently reject art or information forms lacking high levels of sensual stimulation (that is, high degrees of retinal or tympanic stimulation). It is obvious that conceptual paths are tied to fading print literacy and ideas (theory), perhaps the domain of a print-literate elite holding onto an abstract, arcane concept of self, exclusiveness, and privacy. From an opposing sensualist's perspective, any reduction or self-limitation in volume or fidelity of sensory complexity is an incomprehensible move toward sensory deprivation.

Option 2 — The Multisensory Approach: Digital multimedia has potential as a return to perceptualism, as a multisensory, immediate, interactive rejection of the conceptual and theoretical path. Ultimately, digital multimedia is going to run parallel to analogical perceptualism (postliterate or postconceptual awareness). Virtual reality, as the potential zenith of digital multimedia, will amount to the detailed construction of a realistic second nature.

This option is the sensual approach. The strategy is to seduce the viewer-listener with the richness of how things look, sound, and feel, preferably simultaneously. If the audience is challenged with complex sensory information in high fidelity, ideas or concepts are superfluous and unnecessary. Ultimately, digital multimedia will run parallel to and in sync with the phenomenon of analogical perceptualism (the post-print-literate or post-conceptual awareness prevalent in young audiences today). Virtual reality, as the potential zenith of digital multimedia, will amount to the detailed construction of a realistic second nature. VR promises to eventually deliver an environment of total perceptual seduction.

In recent years a number of artists have chosen to take the nonretinal approach. Visual and media artists have migrated into radio, the Internet, the World Wide Web, and any number of immaterial and physical public spaces; the cultural hinterlands. Even television has this potential (as it continues to disintegrate and feather out into more and more specialized channels and Web addresses). Artists disperse from the traditional art zones and seek the resources and, more importantly, the new contexts they need to survive and prosper. History is spatial and psychological, and this evasion of the pigeonhole of art has been occurring for some time.

When artists find themselves in these wide-open territories, they meet musicians, composers, theoreticians, writers, performers, disenfranchised bureaucrats, independent curators, and audiences with fewer preconceptions and far less restrictive cultural filters.

> > >

Making Things Simple (and Personal)

Why do visual artists and visually biased media artists find the medium of radio so attractive? In this era when advertising executives affectionately refer to audiences as numerical sets of eyeballs, in the visual and media arts we see filmmakers, video artists, and computer animators migrating more and more frequently into predominantly or even exclusively sonic zones of operation. What are the benefits of this auto-amputation of sight and abandonment of visual representation?

The answer to this question, for me, begins back in 1974, when I was a young artist living and working in Toronto. I was conducting an investigation into mid-twentieth-century research into sensory and perceptual deprivation. Sensory and perceptual deprivation research was all the rage back in the fifties and sixties. It was a combination of a paranoid, Cold War fascination with brainwashing following World War II, the Korean War, and the energized period of time surrounding the development of perceptual psychology. This period marked the time when a scientific, behaviourist approach was taken toward understanding new media and their effects.

My personal investigation was driven by a simple problem. I had taught myself to write by writing explicit, visually descriptive prose. I wrote hundreds of short texts, all characterizing absolutely silent acts of visual perception. The silence inherent in this visually biased work made me realize that for me there was no connection between writing and speaking. I had no sense of my voice when I wrote. I sat in the dark and listened — and there was nothing. I could not write with my inner ear. Writing for me was entirely a visual experience.

This realization led me to experiment with earplugs and soundproof earmuffs. But there was still not the faintest whisper. Nothing worked until I started experimenting with a *ganzfeld*. A *ganzfeld* is an even, monochromatic field of light. If you take an ordinary ping-pong ball and cut it in half and place these two halves over the eyes, and then look into a bright source of light from a lamp or window, you load your visual perceptual system with an even source of white light. There are no objects to be seen. There are no figure-ground relationships. Light pours into eyes unable to see in conventional terms. In this state, I was first able to detect the weakest presence of my inner voice. I may have just imagined it at first. But after several attempts, I found my voice and I could finally hear my thoughts come to life. I remember how I felt whole for the first time.

Strategies for Recovery

In this age of digital multimedia, human perception is often challenged by dense, multilayered audio-visual-data information displays often more perceptually demanding than the two predominant media environments of our time, cinema and television. This is certainly true of the multisensory approach toward making art. The density and power of simultaneous, multisensory messages is complicated further by demands for interaction and strategies for environmental integration (installation). In other words, the multisensory approach seeks to fill (or overload) the human perceptual apparatus while integrating the audience with the work (interactivity) and in connecting the work (and audience) with the immediate environment (installation).

My research into sensory or perceptual deprivation was, of course, a reductive approach, more akin to the nonretinal approach to art making. By reducing stimuli or closing off much of what I normally perceive of the world — or more specifically by letting light without external, discernable images into my eyes and listening to silence—I was able to hear just a whisper of my inner voice, of my commentary, my thoughts, my ideas.

By Eliminating Visual Image I Eluded the Point-to-Point Logic of a Tightly Bound Rational Constructed Space and Found New, Open-Ended Perspectives in the Associative, Analogic, Transrational Territory of Auditory Space

Within this unusual perceptual state (vision absent and sound absent), things were fresh and new and were naturally horizontal or lateral, clearly emphasizing the possibilities at the periphery. One thought drifted into another, and I was surprised by the lack of order, direction, or logic. It was as if the voice that translated my thoughts into sound was being fuelled by the energy of the neutral light. As I searched for form in the spatial void of the *ganzfeld*, and my idling visual system was oscillating with futile attempts to establish figure-ground stability, I simply noticed that I was thinking differently, not being governed by the point-to-point order of my normal sight.

Basically, I was escaping my visual bias. I had been perceiving the world as the human equivalent of the visual art gallery or museum. I was, in effect, perceptually malformed, the way these visual art spaces are perverse with their clinical white walls and their deafening silence. People whisper when they look at visual art in a gallery or museum, as if they were in church. Radio is the antithesis of the silent visual art space. Radio is a strange zone where there are no pictures, no visuals. Of course, we mix our radio with life, when we listen in our kitchens or when we are driving to work or later while we listen at work. Galleries and museums rarely mix life with their silence. If you want a demonstration of what I'm spelling out, just take your favourite radio program into a gallery with a boom-box-a-blaring and see how people react.

Somehow, I had become the human equivalent of an art gallery or a museum. My *ganzfeld* experiments made me realize my perceptual apparatus was seriously unbalanced. It was time to tilt my world the other way, toward my sense of hearing.

Before you decide that my return to this period of rather eccentric personal research is senseless, let me outline some of the research I barely touched on in 1974. Although this early, mid-twentieth-century, perceptual-behaviourist research is clearly treating people like rats, almost humorously so, I'm reinvesting in this obsessively clinical scientific literature because of its now almost refreshing biophysical bias. Humans were looked at as animals back then; animals that were bioelectrical receivers of symbols, signals, and noise.

The Human Being as a Primary Target for the Reception and Storage of Information

In the past fifty years, psychology seems to have evolved from a study of rats to a study of computers. But let's pretend that the human mind has always been and remains central to the study of psychology. And to help make this fantasy vibrant in 1997–98, I'm looking back, viewing this mid-century clinical research as the rough sketches of a schematic diagram of the human being as a primary target for the reception and storage of information. In any case, I found the literature far richer than I remembered.

The Research Itself

From *Sensory Deprivation: Fifteen Years of Research*, John P. Zubek (ed.), Appleton-Century Crofts (New York, 1969): Fantasies were about four times as frequently reported in conditions of "vision absent/sound present" (versus "vision present/sound absent").[1]

In these studies "vision absent" was accomplished with *ganzfeld* goggles (providing invariate visual stimulus). "Sound present" conditions were created with headphones issuing white noise with varying drumbeats mixed in. "Sound absent" conditions were an even flow of white noise. "Vision present" conditions were simply goggles with clear lenses, permitting sight of the immediate environment.

In other words, tipping the sensory balance toward the ear will likely stimulate fantasies. Where I did find I could easily induce vivid daydreaming within the silent state of my *ganzfeld*, the greater gain in retrospect was finding a sense of inner voice and a connection between thinking and speaking and forming an analogous, lateral, nonlinear, paralogical, postrational state of consciousness.

Once my inner voice became stronger (fully audible), I added instrumental music, improvisational jazz, and all manner of experimental, highly variate auditory input to my *ganzfeld*.

Were Creative People More Likely to Have Fantasies?

"The results suggest that creativity is strongly related to information drive in high-creative subjects, but does so only after a considerably longer period in low-creative subjects, whose 'chronic' level of information drive in the extra experimental environment is presumably at a correspondingly lower level".[2]

Put another way, more highly creative people have a greater "information drive" and get into spontaneous, internal image and sound generation (fantasy) more quickly than less creative types. It was then stipulated that "chronic" information drive was flattened by decreasing levels of stimulus variation.[3]

There was also neurophysiological evidence that the nervous system is in constant activity and not passively waiting for arousal by stimuli. The activity of the brain itself constitutes the "drive" of behaviour.[4]

More about the "Information Drive"

The ultimate value of sensory deprivation research is found in its contributions to behavioural theories. The motivational theories of the 1930s placed a relatively great importance on specific physiological drives such as hunger, thirst, sex, and pain avoidance. In this context, the idea of an "information drive" seems only natural.

The information drive seems to be governed by what researchers called "cross-modality effects."[5] Heavy doses of visual information may reduce the need for auditory stimulation and vice versa. And a balance in visual-auditory information drive may be achieved through cross-modal stimulation. "There is strong evidence that the information drive in humans mediates a homeostasis of information transmission, and that the homeostasis, once disturbed by extreme values of stimulus information in a particular modality, may be reinstated, at least in part, by compensatory adjustments of the information value of stimulation in another modality."[6]

But there are also conflicting messages in the interpretation of this research. Perceptual and sensory deprivation research was based on eliminating environmental stimulus as a means of observing the psychophysical functionality of the human subject. Balancing or shifting sensory bias, or refreshing the human information drive through strategic, additive means is another story. These researchers did not attempt to formulate the cultural effects of actively alternating or weighting the modality of visual-auditory stimulus or changing the frequency and volume of same. This additive, cross-modal research was left to advertising agencies, politicians, artists, musicians, and the entertainment industry.

Generally, sensory-perceptual-information deprivation researchers believed that the effects of visual and auditory information satiation on information drive were complex and varied greatly from individual to individual. Most of the time, visual information satiation would lead to an increased information drive for auditory stimulus and vice versa. It was generally believed that visual and/or auditory information satiation, if the stimulus was invariate (unchanging), would result in a suppression of the information drive.[7]

Could This Clinical Research Be Useful in Gaining a Critical Perspective on a Contemporary Information Environment Impoverished Due to Its Invariate Physical and Symbolic Nature?

The literature I am describing is from the arid world of mid-century clinical psychology. The authors are researchers probably depicted best in white lab coats. The symbolic nature of information during this period is best epitomized as the green squiggly electronic line representing the vibrations of speech on an oscilloscope. Television back then was exclusively black and white. Typically, there were only three channels of talking heads, a constant flow of cigarette and beer commercials, automobiles with tail fins, tiny cowboys riding from one side of the screen to the other, and the occasional fantasy about outer space. We had not yet walked on the moon. The Western culture was minimal and uncluttered. Compared with today's vibrant, densely layered, fully loaded hyperculture, the 1950s and early 1960s were like one huge sensory deprivation experiment. So where's the connection between this literature and today's reality?

Perhaps the key to making this jump is the abstract and highly portable idea of invariate stimulus, regardless of its physical or cultural dimensions. Sure, we have jettisoned the old fuzzy black and white world as we have adopted the wraparound, full-fidelity frenzy of today's electronic culture — the pulsing, throbbing, shrieking barrage of constantly refreshed streams of symbolic code. Today, everything is always *new*, always changing, and paradoxically always the *same*. Everything is bright and shiny and jarring. As creatures who feed on information, a diet of the hypersame makes us all poorer for our efforts — bloated but empty.

Could Confinement in an Impoverished Media Environment Generate Boredom and Monotony?

The symptoms of our information-feeding disorders are beginning to define a culture of human inadequacy. Strangely enough, there were signs of our inherent weaknesses in some of those research papers back in the 1950s.

In *Sensory Restriction: Effects on Behaviour*, by Duane P. Schultz, Academic Press (New York, London), 1965, Schultz reports that in 1957 a scientist named Cohen was working with a coloured *ganzfeld* that "utilized two intersecting spheres, both illuminated by a highly saturated red light. On first looking into the apparatus the subject sees a poorly saturated red fog. Shortly this color begins to fade and appears achromatic in about three minutes. This disappearance of color in a Ganzfeld demonstrates the vital role played by spatial inhomogeneity and temporal change in perception." [8]

"In another study restricting stimulus input to homogeneous visual stimulation, Cohen (1958) demonstrated the 'white out' phenomenon (cessation of vision) under both monocular and binocular conditions . . . The 'white out' phenomenon tended to be suppressed by factors which introduced sensory change such as blinking and eye movement and the presence of an object in the visual field." [9]

Could This Be a Precursor of Snow Crash or Blanking?

From *The Brain Benders: A Study of the Effects of Isolation*, Charles A. Brownfield, Exposition Press (New York, 1972; this second, enlarged edition was originally published as *Isolation: Clinical and Experimental Approaches*, 1965): There are "four general categories of isolation: confinement, separation, removal, and monotonizing of stimulation." [10]

"Social isolation entails only the elimination or reduction of communication through the senses of one individual or group with other individuals or groups." [11]

"At this point it should be apparent that human beings are individually, socially, and physiologically dependent not only upon stimulation per se, but upon a continually varied and changing sensory stimulation in order to maintain normal, intelligent co-ordinated, adaptive behaviour and mental functioning." [12]

In rereading Brownfield's research findings, it is difficult not to form the broadest analogies about how individuals are affected by invariate stimuli and monotony. Just substitute a 1990s cultural environment, overloaded with amplitude and density (but underloaded in terms of invariate messages), for the tightly controlled clinically impoverished environments of mid-century sensory and perceptual deprivation. Is it possible that we have accomplished the same levels of environmental impoverishment and isolation by additive rather than reductive means? Could we be witnessing widespread feelings of isolation and human inadequacy as a result of our loud, dense, largely repetitive, redundant, machine-induced electronic cultures?

Youth Argue for More Noise

Today's youth would argue for more noise and greater and greater perceptual challenges characterized by density and complexity. This is, of course, the traditional position of all youth now and in the past. The need for and tolerance of optimum levels of stimulation peak in adolescence.[13] High school (fourteen- to eighteen-year-olds) and traditional college-aged students (eighteen to twenty-three) exhibit chronically low information drives, often appearing comatose during school hours, or when receiving information in single sensory modes (monotone lectures without audio*visual* aids = "vision absent/sound absent"). Their perceptual systems need quite a kick to stimulate optimum performance. This is also the case with adolescent and postadolescent youth who drop out of school, taking minimum wage jobs or joining the ranks of the unemployed. It can be argued that these dropouts inherently have extremely low information drives and crave even higher volume and density from cultural sources, when their perceptual systems come alive.

Typically, adolescents awaken slowly (morning classes are not productive), and conduct their entire school days in states of low arousal. In the evenings, they pass through their optimal performance periods immersed in multisensory stimulation (sometimes doing homework with their music or television in the foreground) until they experience overload and perceptual disorganization resulting in emotional disturbance. Finally, they crash. Their sleep, an escape from exhaustion, appears to be deep, but it is typified by high frequency, low voltage brain waves. It is a sleep without real rest. If they achieve delta sleep at all, sleep characterized by low frequency, high voltage brain activity, it is just as their alarm clocks go off. It is time to enter the numbing fog of another school or work day. If adolescents need and can tolerate a culture of high sensory impact, younger children and adults need and seek protection from the same. A three-year-old cries in terror when his or her balloon breaks. Recently, in the dark spaces of a Bill Viola retrospective at the Whitney Museum of American Art in New York, I noticed that many children (four- to eight-year-olds) were completely intimidated by Viola's video installations. The "fun house" environments (dark rooms, loud audio, big bright video projections) were simply too much; they were too frightening for these kids.

The middle aged and elderly are not frightened, but they typically are plainly irritated by high amplitude and multisensory density and complexity characterized by today's youth culture. Noise ordinances are designed to protect adults and young children from the excesses of youth culture, and, of course, from the assault of general, all-pervasive noise pollution. Multisensory art often occupies the territory between youth culture and levels of noise typical of urban or industrial disturbances. I must add that the middle aged and elderly are commonly annoyed by high frequencies of image change and clearly assaultive or violent images, fearing that this reactive muse may be interpreted as a rant against strictly auditory attacks on the human perceptual system. It is more important to recognize that most of our entertainment culture is operating on the fundamental idea that it is good to "fill up" and overload the senses — to give the consumer their money's worth. The multisensory approach to making art often employs this "fill them up" and "overload them" tactic. Although adolescent audiences never seem to get enough, young children, the middle aged, and the elderly quickly suffer from overload.

Coping within Machine-Based Cultures

If it can be argued that invariate cultural environments can cause boredom through monotony and that "information drive" in a wide spectrum of society is suppressed by the same conditions, is digital multimedia with its promise of full immersion and engaging interaction really the answer? Amplification, density, supranatural fidelity, and complexity may arouse for brief periods, but these aspects of machine-based culture and art will, of course, continue to numb down the last organic components in our home-entertainment systems: namely, us.

The advance of machine-based cultures is not isolated or strictly segregated from more organic cultural phenomena. Machine culture is carried into the world via portable systems, be they traditional like audio systems in automobiles, or more recent developments like boom-boxes, walkmen, and disc-men, or video camcorders with LCD screens and external speakers, and all manner of wireless telecommunications devices and wearable computing . . . This intrusive superimposition of artificial, supranaturally loud, dense and complex electronic culture over the "natural" environment is a spillover both contaminating our streets and creating an addiction to noise. Increased levels of amplitude, complexity, and control are necessary for sufficient perceptual arousal.

Whether you are harassed by the piercing electro-acoustic assault on the street or you are blissfully experiencing the cinematic-like reality induced by a disc-man's clean sound in a set of full-cup headphones, both the harassed and the seduced eventually share a sense of exhaustion. The cost of the unwanted or desired hyperperceptual arousal is much the same at the end of the day. We break down and experience a feeling of human inadequacy. For the harassed, it is anger and fatigue in response to the assault ("I can't take it any longer" "I can't think straight"). From the position of the seduced, those deriving pleasure from an immersion in their own electronic polyphony, things are different, but strangely similar in emotional tone ("I can't get enough" "I can't pay attention to anything anymore").

To double back to the beginning of this text, I stated there were two primary, polarized, oppositional strategies for making contemporary art within machine cultures: *Option 1 — the Nonretinal Approach*, the reductionist path of eliminating sensuality by limiting the multisensory spectrum of the work, and *Option 2 — the Multisensory Approach*, digital multimedia, the additive approach of filling up the senses. I have struck such a stark and elementary, dichotomous model for a reason. In reality, our machine cultures are so dense and disorderly in structure they are almost impossible to comprehend. The real hybrid currents of culture and art are oscillating wildly back and forth between the sensual and reductionist poles of this model. In fact, an analysis of contemporary cultures and art will demonstrate permutations and hybridization galore, as reductionist positions are surrendered to sensuality and vice versa, ad infinitum. In this electronic or digital era, so much of this frenzy of hybridization is dictated by machines with their unprecedented potential for manipulation and control, new levels of fidelity in representation and fabrication, and the power of unfettered network distribution.

Sensuality versus Deprivation

Where I have obviously benefited personally from research and experimentation within the reductionist or nonretinal approach — I've found my voice and managed to establish a new freedom in how I think and make art — I have also pointed out that the multisensory approach is considered by most to be sexier and will likely hold the upper hand from here on out for good reason. To get any real attention in machine cultures, you have to be more overtly demanding and aggressively manipulative than everyone else. That means amplitude, frequency, intensity, and, of course, inescapable interaction and full immersion. In the noisy cultural context I have been elaborating in this text, stripping things down and "going quiet" or arcanely incomplete (alphanumeric codes: text, ideas, concepts, sound without image, the naked voice) seems shortsighted and ultimately futile. Reducing direct sensuality makes things symbolic and indirect.

To be blunt, taking the reductionist position will be akin to depriving one's audience of "normal" or healthy levels of sensory arousal. The reductionist strategy will be labelled "deprivationism," and criticized as such.

Perhaps to the disappointment of all, I conclude that just as in life, nothing is simply black and white—and neither of these polarized strategies will yield satisfying results in the short or long term (twenty-five to a hundred years into the future). In fact, I infer that additive and reductive approaches will yield strangely similar results, as long as the primary target for the reception and storage of information, culture, or art, is the human mind (machine culture and art may be more constructively aimed at machines, not people). As humans, we are simply too physically and culturally exhausted at the end of the twentieth century to express our preferences or reactions clearly.

Machine Cultures Are Increasingly Designed to Engage Adolescent Perceptual Systems

I link physical exhaustion with cultural concerns because it seems clear that we are creating machine cultures that treat whole populations as adolescents. Machine cultures are designed to stimulate those with chronically low information drives and like all good technological solutions, they compound the problems they are designed to fix. Injecting more noise and intensity into an audience that cannot respond because of fatigue and inadequacy is guaranteed to induce exhaustion, confusion, and self-doubt. There will be more and more gaps or implosions of reality where consciousness and meaning break down and disintegrate. The human perceptual system will progressively fail to negotiate the extreme, posthuman realities dictated by machine-based cultures. As punchy as we are as a species, both overload and underload will lead necessarily to anxiety, emotional disturbances, perceptual disorganization, and disorientation. In short, only one thing is certain: we will increasingly experience feelings of human inadequacy as our machine-cultures continue to advance.

> > >

Addendum: Trouble Signs

Blanking (Overload): Increasingly, we "blank" as we try to negotiate environments where multiple sources of information are mixed or blended. Multiplexing or multitasking are machine traits we wholeheartedly embrace with mixed results. We love to extend ourselves but are certain to crash.

Spacing Out (Underload): Our attention or focus drifts when we encounter thin or low-impact information environments, such as music without images, people just talking in their inherent monotones, or alphanumerically encoded information such as math and literature.

Pumped but Capped (Information Toxicity): In cultures where expression is limited to those with access to the skills and tools of production and means of potentially dialogic distribution (networks), active creative output is tragically blocked, and people simply clog up and shut down, if they survive at all.

Pumped Dry (Inputting into a Vacuum): In machine-based cultures, those with the skills and means to produce and distribute cultural information often pump themselves dry, burning themselves out trying to match the scale and potential of machine-driven opportunities with their organically limited energy and efficiency.

Empty Language Yields Empty Experience: Words like information, which may be the most significantly empty word ever to find widespread use, diminish the descriptive power of language and in effect reduce our understanding of the world. The expanding blank space of "information" in our language is not only increasingly without shape and meaning in abstract terms, but as such empty words are applied to sublanguage realities, they clearly serve to eradicate the defining characteristics of our experiences in the flesh.

1 Zubek, 52.

2 Zubek, 192.

3 Zubek, 199.

4 Zubek, 408.

5 Zubek, 200.

6 Zubek, 201.

7 Zubek, 200, 201, 408.

8 Schultz, 84.

9 Schultz, 84.

10 Brownfield, 10–11.

11 Brownfield, 70.

12 Brownfield, 74.

13 Zubek, 430.

The Finished Work of Art Is a Thing of the Past 1995

At the close of this century, we are witnessing a major change in how value is determined. The value of material wealth is giving way to the value of information. In this time of transition, these apparently incongruous value systems mix and form hybrid systems for determining value. Scarcity, even in an era marked by an abundance of information, is still a key factor in determining value. Unique, precious material objects still hold their value (in an economy not unlike one originally pinned on the gold standard); some actually increase in value in a relatively short time. Information that is useful, but scarce, is also valuable. Those who hold valuable information may still wish to maintain exclusive, proprietary control—to increase the life of the information. Information is subject to decay or aging. Information is not inexhaustible—it may revert to data, the raw material from which it is formed. How and when information is maintained and released is determined by those in control: those who initially recognize its value as information and manage it and operate with it accordingly.

Contemporary art is part of an emerging sector of the economy called information and knowledge. Knowledge workers are people who create information for others to use. "Worker" in this case does not imply blue-collar or white-collar worker; that is, those who act only upon the instructions of others. Knowledge workers think for themselves. They know things that other people do not know. They can solve problems and help others solve problems. Knowledge workers produce information; they create information by taking incomprehensible data and transforming it into information—distinguishing key aspects of disorder with form. Artists fit nicely into this role of knowledge worker. Contemporary artists, curators, critics, and art historians are the knowledge workers who form the contemporary art domain of the new sector of the economy called information and knowledge.

Contemporary Art Has to Be Seen as Information to Be of Value

The product of the knowledge worker is information. The product of the artist is art. In an economy where value is determined by information and knowledge, art has to be seen as information to be of value. The process of creating information requires a set of skills, methods developed through higher education, and experience. Creative processes, in their most basic forms, can be taught and learned. Creative processes thus become products in themselves. Although these creative processes have universal characteristics, creative work is messy, chaotic, and mysterious by nature. In art schools, experienced artists attempt to teach young, emerging artists how to work, how to create art. It is difficult to teach and learn creative processes. Most art and information resulting from the study of creative processes is of little use to an audience beyond the initial teacher and students. The exchange of creative processes is even more difficult outside of the academy. Most art or information resulting from creative processes is of little use to other people, whether these other people are creative or not.

Audiences for art, the consumers of art and creative processes, must aspire to be creative themselves to find the products of artists valuable. f the work of art is an object, then an audience has to be able to decode the material object to extract information encoded in the object itself. This participatory investment, an interaction between the audience and the object of art, is most commonly described as interpretation. It is an intuitive, subjective process verging on psychic identification. Essentially, the work of art is performed by the audience, permitting the consumer of the object of art to retrace the creative processes of the artist. In this manner, the consumer is involved in a kind of virtual creative process. With contemporary art, the curator and critic offer their assistance in this information exchange. Dealers and artists themselves also go to great lengths to help audiences find ways of relating to objects of art. The dealer loves and respects his or her artists; the artists display their personalities in public appearances, offering clues for interpreting (identifying with) their works. Galleries and museums focus the attention of audiences on objects of art in isolation: the "white cube" removes the artwork from the world at large and permits the work's aura to be witnessed in a quiet contemplative serenity; that is, conditions conducive to psychic identification.

The object of art described above is a material manifestation of vision and thought; the artist sees and thinks and completes the creative process by fixing his or her perceptions and experience in material form. Traditional forms, painting and sculpture, have been extended through photography and architectural manipulations (installation), including attempts to integrate information technology (video and computer-integrated media). Over time, there have been attempts to shake up the system of material, object-based art. Ideas have become accepted as art (conceptual and neo-conceptual art) and performance art and other kinds of "live" events (often via telecommunications) have been introduced as art; performance art being centred on the physical reality of the body itself. These purely conceptual and "live" forms have found more receptive audiences in public spaces, spaces not defined as art spaces, such as museums and galleries. When artists move into public spaces, they find that audiences in these neutral spaces invest in their work differently. They are not interested in history, not even in recent history. They are interested in today: the experience at the moment of exposure.

The problem with conceptual or live forms has been how can one exchange art to economic advantage if there is no adequate system of currency, an accepted symbolic medium of exchange? Part of this economic problem is due to the fleeting, immaterial, and impermanent nature of conceptual and live forms. The work does not manifest itself as a currency for exchange. There is no accumulation of material history. This has led to an obsession with documentation (catalogues, interviews, photography, video, CDs, CD-ROMs) and the indirect commodification of live forms through mementos or souvenirs (limited edition prints or drawings or other unique material objects of art). A solution to this problem is apparent in the recent proliferation of immaterial objects. Open-ended, "living objects," or more precisely "living systems" are now being produced under the generic description of software. Living systems, such as expert systems or other manifestations of artificial intelligence-perception-experience, while in their primitive stages of development, will soon present a serious challenge to fixed or "finished" art as we know it today. Artists working in live forms will first collaborate with the experts who make immaterial objects (multimedia programmers), then they will learn how to author these living systems themselves.

The Living Artist's Body of Work is Never "Finished"

Let us return to what we understand fully. When we visit a museum or gallery and view objects of art, a key aspect of the interpretative process is based on knowing when and where the object was made. Artworks by living artists are viewed differently than works by deceased artists. Curators, critics, art historians, dealers, and artists will always point out the importance of knowing an artist's entire body of work when attempting to decode a single, discrete object of art. When an artist dies, his or her body of work is completed — it is finished and each object of art is a fixed component of a body of work, complete in and of itself. The living artist's body of work is an open-ended, expanding work-in-progress, and therefore each single, discrete object of art is part of the body of the unfinished work-in-progress as well. An audience interpreting such unfinished work must update the work with more rigour than they update the "finished" works of deceased artists. The work of dead artists can be decoded for the information it provides about a specific period of time in the past. Although the works of certain deceased artists sustain their value as information and therefore increase in value as material objects, this value is based in their concreteness, or ironically the relative fragility of their finite material reality. A painting from the seventeenth century can afford to look dated. Audiences are not so kind when viewing contemporary art by living artists.

Information Is the Perception or Experience of Difference

Contemporary works of art are valued most if they appear to be up-to-date. The fresh "new look" always has value. Hot new work by young artists, or brand new twists in new works by established artists, have the look or appearance of information. New technologies are great for achieving the "new look" and for actually creating the "look of information," whether the work achieves value as information or not. Contemporary artists, young and established, compete for the most up-to-date look. Appearance is a territory. Curators, critics, art historians, dealers, and artists participate in this search for contemporary artists who distinguish themselves as being hip (informed) through the look of their work. Audiences are exposed to the newest, most informed work, and they verify through their excitement and energy if the work is charged with information value. The new look can never be predictable. Information is always a perception and an experience of difference. It must come as a surprise. The most unlikely things slam together to become information. Often it is impossible to explain why specific works are so information-rich. Information is fleeting. All information is time sensitive, and the hottest new things age very quickly. In a very short time, the new look becomes tired and old. A work by a living artist that has gone out of fashion is practically lifeless (95 per cent dead).

The "new look" can be easily dismissed as the primary attribute of superficial art, shallow art that will quickly fall out of fashion. Surely, an artwork of real depth will stand the test of time. Deep works are structured so their information is released slowly, over the long run. Usually, these deep works are seen to function as universal knowledge structures. Universal works of art defy identification with a specific time or place. From another perspective, perhaps the deepest works of art are constructed to be totally devoid of information, functioning as attractors of layer upon layer of incomplete interpretation, an ongoing investment of intellect? Such speculation on how a particular work of art will function in terms of information can be addressed by asking three questions about any work of art: Is the work of art loaded with information? Is the work of art totally devoid of information? Does the work of art transform data into information?

The first question asks if the audience is informed by the work. Does the work offer a rich field of information and knowledge to the viewer? Can the viewer learn and take something useful away from the experience of the work? The second question asks if the work of art defies those who would try to extract information from it. A work of art totally devoid of information could be labelled anti-information. It offers no information and literally rejects all associations with information. The third question reintroduces the term data, the raw substance from which information is created. Artists as knowledge workers distinguish aspects of disorder, previously indistinguishable data, with form. They perceive difference, significant difference, and construct situations (frequently in or through the use of material objects) where information can be produced *by the audience.* Does the work of art transform the data around it, a data field that is constantly changing in time, into information?

These three questions can be asked of both material and immaterial works. In a digital era, immaterial "objects" are increasingly prevalent, but their function will initially differ little from their material predecessors. They are carriers or rejecters of information, or transformers of data into information. They differ in that immaterial objects of art function better in virtual spaces. Where museums and galleries are physical architectural sites best suited for the public presentation of material objects of art, immaterial objects function best on networks, virtual public spaces that connect private spaces. At the turn of the century, there will be a continuing, increasingly more complete synthesis of material and immaterial objects, of physical and virtual architectures, of value systems based on dichotomies of scarcity and abundance, material wealth and information wealth. Mixed economies, perceptions, and experience will countervail to produce radical new degrees of hybridization.

Identity Is Address in the Territory of Experience

One single critical problem emerges from this period of chaotic transition. How can a work of art be updated so it does not lose its value in such a volatile information environment? Information (and certainly art in an information age) has a very short life. Contemporary art begins to fade immediately after it is exposed to an audience. The living artist updates his or her body of work through subsequent releases of new work. The living body of work is continually updated through twists and turns, rarely complete flips or radical changes in direction. The thread of consistency, aesthetic logic, must remain unbroken. At the core of the living artist's evolving work, there must be a redundancy of form and a consistency in the method by which disorder is processed into form. This redundancy of form and method must create enough consistency so the work has a recognizable identity. This identity is the address of the work—the site of information. An evolving body of work has no fixed address except its recognizable appearance. Identity is address in the territory of appearance.

All mature artists with considerable bodies of work eventually fall victim to the weight of their own history. Each discrete "finished" work remains stable in a contemporary information environment characterized by great volatility. The stability created by one's own history (a body of finished artworks) produces the artist's address in the spaceless territory of appearance: his or her identity. This address, while necessary for recognition, unfortunately dictates the parameters of new works. Mature artists must continuously update what they do while remaining consistent with their past work. In fact, a common strategy in the twentieth century has been to remain completely consistent — to repeat the same information as art over and over and over. In a twenty-first-century culture, value will be determined increasingly by the freshness or newness of information, and this vitality of information will be based on the timely, continuous introduction of apparent new qualities. Stability, unless it is poetic ("Poetry is news that stays news," said Ezra Pound), will not be a positive attribute in an environment characterized by continuous, constant change. The difference between poetic stability and mundane redundancy is that poetic information is volatile in its own way.

Ethereal Culture Is Grounded, Made Concrete, in Its Audience

A stable, poetic information structure (let us call it a work of art of an ambiguous nature) yields information through the creative efforts of its audience.

The audience interprets (participates, interacts with) the work to create information. The chief edge that immaterial objects have over material objects is their potential for direct, active participation and interactive manipulation by audiences. Immaterial objects fly back and forth across networks at the speed of light into private spaces where audiences can manipulate and modify identical, digital copies of original artworks, updating these works as their information is consumed or, more correctly, processed. This is why interactivity is such an obsession in computer-integrated media. The digital-arts sector is completely chaotic and volatile except for the potential of societal integration via networks and connectivity. Ethereal culture is grounded, made concrete, in its audience. The identity or address of the work is shared by the artist *and the audience.* Works of art in the immaterial domain are never finished; they are simply introduced (initialized) and placed (contextualized) for the audience to participate-interact with: to add to, change, customize, work on, pass on, and more. The artist, of course, may choose to revisit any or all of his or her previously finished works for revision in such an interactive environment.

This updating process, where the artist sends instructions for transforming his or her entire existing body of work by adding, subtracting ... emphasizing, amplifying ... twisting, tweaking, reversing, transposing ... recontextualizing, destroying ... (you name it), is how attractiveness, vitality, and ultimately value will be maintained and recreated in the new information economy. The artist will no longer simply be as good as his or her latest work—the artist's work will only be as valuable as it is up-to-date. Works of art that were previously valued because they represented concretely the perception and experience of a particular, fixed period of time, will have to be updated and at least partially reformed to maintain their value as information today. The artist, while living, will participate in this updating process with his or her audience: curators, critics, art historians, and everyone else who has access to the work.

Radio and Television and Telecommunications-Based Multimedia Continue to Undermine "Finished" Art

For the first seventy years that art has coexisted with instantaneously updatable electronic media (radio, television, and most recently telecommunications-based multimedia), artists have functioned primarily as reactionary figures, producing material history valued largely because of its stability and strong ties to the past. Things are far more complex in the 1990s. Today, we see artists moving in two completely opposite directions, determining and fulfilling the criteria of two incongruous value systems. For some, value is determined by establishing and dating "finished" works of art—setting up stable structures designed to hold and perhaps increase their value as they recede into the past. For others, art is valuable only if it is current—if it is a living system, characteristically fluid or "liquid" in nature. Building in mechanisms for updating works of art, such as interactive mechanisms, is a strategy for maintaining and recreating the value of art *as information*. In such interactive works, the potential interval frequency of revision will become the primary factor for determining value.

Between these divergent value systems exists a broad spectrum of hybrids, paradoxically including traditional material objects that function as information generators of the highest order of sophistication. And, of course, there are also immaterial objects structured as "poetry," which stand rock solid against the swirling, chaotic patterns of change. Art that produces information in its relationships with audience, transforming data into information, is always a living system (whether concrete and solid, fluid and liquid, or fleeting and ethereal). The contemporary artist and his or her audience breathe life into such living systems or, more precisely each distinct species of living system. The evolving living works of living artists naturally defy completion and stasis and death (and extinction ultimately—being discarded and forgotten). An emerging strategy for survival is building in mechanisms for updating. Making interactive works that are updated as they are used has potential as a survival strategy.

Programming interactivity is all about the potential for sustaining the life of the work; making updatable art invites the audience for the work, including the living artist, to participate in the creation of the future of the work. This sustained life of the work is in contrast to the usual abandonment of the "finished" work of art. As we look back at those "finished" works of art, left behind to sustain themselves (in special vaults, the "white cubes") against the ravages of time, it is clear that the "finished" work of art is a thing of the past.

> > >

Bibliography

Bell, Daniel. *The Coming of Post-Industrial Society: A Venture in Social Forecasting.* New York: Basic Books, 1973.

Drucker, Peter F. *Post-Capitalist Society.* New York: HarperCollins Publishers, 1993.

Feather, Frank. *G-Forces: Reinventing the World, the 35 Global Forces Restructuring Our Future.* Toronto: Summerhill Press, 1989.

Gelernter, David. *Mirror Worlds: Or the Day Software Puts the Universe in a Shoebox ... How It Will Happen and What It Will Mean.* New York: Oxford University Press, 1992.

Hardison, O.B. *Disappearing through the Skylight: Culture and Technology in the Twentieth Century.* New York: Viking Penguin Books, 1989.

Nora, Simon, and Alain Minc. *The Computerization of Society: A Report to the President of France.* Cambridge, Mass.: MIT Press, 1980.

Penzias, Arno. *Ideas and Information.* New York: Touchstone/Simon and Schuster, 1989.

Poster, Mark. *The Mode of Information: Poststructuralism and Social Context.* Chicago: University of Chicago Press, 1990.

Museums of Tomorrow 1987

Your self-righteousness comes from where you reside.
– Anonymous, January 1987

Museums are places where dead things are kept to be remembered. Art museums are architectural vessels that preserve representative works of the past. If artists themselves and not just their works were buried in museums, then these architectural vessels of memory might be perceived as rather exclusive cemeteries.[1]

In developed countries, it is extremely difficult to secure a plot in an art museum of great stature. In developing countries, art museums are insignificant institutions if they exist at all. Artists in developing countries must align themselves with social and political change if they hope to be recognized. Any country that sets out to construct museums of great stature represents a nation with a death wish of considerable weight. The message is literally set in reinforced concrete. What this country needs is more dead artists.[2]

Let me double-back for a moment to qualify some of the ideas in these introductory paragraphs. The developed versus developing country scenario may be tragically short-sighted in an article speculating on cultural futures. I will be disappointed if it is not. Developed countries will certainly continue to develop, while developing countries will more than likely cease to advance. Many authorities feel that inequities between the information rich and information poor will continue to increase before they begin to diminish.[3]

The problem with measuring the cultural sophistication of a society is perpetuated by disparate, shifting criteria for establishing value. It is safe to say that there are no universal benchmarks to aid in determining comparative states of development of international culture and art. The old standby, aesthetic philosophy, remains virtually ungrounded in general theory. On a more specific, though ponderous note, social statistics and economic indicators are regularly employed to substantiate the rhetoric of the cultural industries. The socioeconomic approach manages to produce graphs and charts that truss up anachronistic ideas and initiatives to the degree that they leverage millions in government grants and corporate sponsorships.[4]

All this is to admit that cultural development is a relative concept. Comparative measurement of cultural sophistication is simply impossible. This impossibility is why death, in particular the death of creative people, plays such a fundamental role in societal development. The flowering of cultural prestige through the accrual of death is absolutely phenomenal if the accomplishments of the deceased are well documented and promoted aggressively. This grim reality is why information technologies and communications media have been able to take their place alongside traditional cultural institutions. More importantly, this reality is why the cultural authorities of developing and developed countries alike are obsessed with death and the new technologies.

Before we entertain a set of ideas comprehensive enough to link the superpowers, the inevitable nuclear wars, strategic defence initiatives, and the silent promise of senseless death with cultural development, let me make myself as transparent as I can afford to be. Please permit me the advantage of representing a few narrow, simplistic, historic caricatures. There is without doubt great cultural stature associated with an historic depth of artistic achievement. Centuries of history enriched with extensive documentation of artistic accomplishment empower the colonialist. Colonialism has always been driven by the grave momentum of snowballing cultural histories. The imperialist, on the other hand, simply dominates with technology and superior distribution systems. The imperialist message may be nothing more than shallow, superficial day-to-day nonsense. The slogan of all imperialist forces is "We deliver."[5]

For a little more than a hundred years, the museum in North America has been the most effective cultural institution for framing history, enabling members of its society to take a good, serious look at the past. Museums have an advantage over other more ubiquitous institutions, such as nonfiction literature and documentary photography, because museums house the images and objects of continuous reality, the physical artifacts that remain intact from previous times. Museums are constructed to highlight and protect collections of such physical antiquities. The value of a museum's collection is based on authenticity and scarcity.

If museums were nation states (and in a way they certainly are), they would likely be perceived as colonialist entities. They are simply too conservative to be labelled imperialist. Colonialist or imperialist, what sister institutions would the museums most likely set out to dominate or plunder? Let them commence with an invasion of the schools. If there is to be a national cultural strategy, then the key to rapid development lies in the creative potential of youth. The schools, and of course the innocent children therein, are extremely vulnerable due to government cutbacks[6] and the disintegration of the nuclear family. The museums could lure more children in on field trips if they could afford to offer more hands-on creative play. Or they could take government money away from the schools for the purpose of paying living artists to enter the schools.[7]

There is a kind of tortuous irony in store for all those who sign up for duty in artist-in-the-school offences. The socially concerned cultural institutions employ artists to eradicate feelings of inadequacy engendered in the student population by inaccessible institutional culture. Museum culture is difficult to grasp because the authorities are committed to raising the standards of art appreciation in a society plagued by the corrosive influences of popular culture (TV, country and rock music, the representation of sex and violence in the media, spectator sports, and lotteries).

Putting aside questions of self-doubt, the artist walks into a classroom as a guest lecturer and is introduced apologetically as a professional creative-type. He or she must contend with a few very serious problems. A living artist has less-than-zero status unless he or she enjoys previously attained notoriety with the classroom audience. With the odds of possible notoriety being 1,000 to 1, the only way out of the hole is a quick demonstration of artistic talent. The right stuff in this context consists of having an uncanny knack for breathing life into art forms ascribed to the deceased. The visual artist must be able to demonstrate hand-eye coordination. Basic drawing skills still distinguish the pros from the cons.

We could dispense with all such painful mention of extension programs if the educational arena was not the most crucial battlefield in the quest for cultural hegemony. The colonialist museums strive methodically, step by step, to clarify the distinctiveness of their cultural mission by introducing an army of living-dead apparently desperate for postmortem recognition by the museum authorities. These shadowy mercenaries drive forward like mating insects into the semiconscious fog of a youth culture drunk and stoned on a relentless image barrage of disposable, instantaneous electronic culture. These half-dead artists in the schools, possessed by the vision of life after death, contribute to the psycho-economic dynamism[8] behind the exploding museum culture.

The museum will continue to grow rapidly in number and stature because as an institution it will provide a public network of technology-free zones within urban centres. There are also more and more collections of things that will need to be housed. Museums are grand architectural sarcophagi, which mushroom into existence for the purpose of protecting and exalting collections of physical artifacts. The fundamental instinct to collect and hoard is rationalized through the disciplines of analysis and classification. Whether the impulse to collect is based on insecurity, altruism, or greed, the economic factors behind every collection correspond to the realities of today's economy at large. Like the current international monetary system, which is tethered (somehow vaguely, historically) to the price of gold, the value of the unique artifact is determined chiefly by factors of scarcity. The ongoing evolution of an information economy, based on the duplication and exchange of free-

floating information, presents a critical assault on the sanctity of such concepts as permanent material value.

Against the incredible tension generated by these conflicting value systems, further institutional growth will depend on aggressive audience development strategies, including the development and implementation of radical innovations such as image-based force fields. Before I undertake to describe an image-based force field, a development which I feel is inevitable, let me explain why art museums move forward so competitively. Art museums are going to face tremendous competition for audience and resources from museums of science, technology, industry, natural sciences, agriculture, civilization, space exploration, informatics, and holography.

There will also be the cinematheques and videotheques, and the technologies like videodisk and digital image, sound and alphanumeric storage media. Databases, now only bibliographic and referential, but spreading like cancer,[9] will quickly develop into virtual environments where computer graphics, digital video, audio, and data can be stored with compositional diversity in micro-electronic space, duplicated repeatedly without degradation and exchanged instantly over ordinary telephone lines.[10] Advancing digital memory technologies will soon be applied to virtually expand physical space.

All of these developing technological realities should offer hope to most living artists tomorrow or the next day, and they will likely foster imperialist fantasies in stolid colonialists in the meantime. The technological vacuum that must be maintained to prevent the collapse of the economy of the precious object will herald an increasing awareness of the technology-free zone. Pockets of virgin nature excluded, the art museum will inherit the exclusive rights to technological vacuity as the church collapses around altars adorned with giant video display screens. The art-museum audience will continue to be monitored for purposes of collection security and psycho-economic research, but all surveillance technology will be completely concealed. Live security personnel will also operate undercover, dressing and acting like museum visitors, so as to alleviate stress incurred while viewing precious objects under constant, overt surveillance.

Image-based force fields will surround the heart of the museum, the technology-free zone, providing an antechamber of unnerving contrast between the contemporary information-rich home and street and the sensory deprivation chambers within the museum. The exhibition of artifacts without the provision of an appropriate viewing context is a waste of time.[11] The technology-free zone must remain uncontaminated if the nineteenth-century mentality is to prevail within the museum. The image-based force field will function mainly as a value rectifier. Icons from the past must be shown to anticipate the artificial, meaningless appearance of today's disposable images and things. The image-based force field will serve to buffer images and objects rendered incompatible by time. Viewer disorientation will be a major factor. Rapid flip-flop change holds the key to value rectification. The frequency of image change should be on the order of 1,800 variations per minute,[12] emanating from multiple, surrounding image sources and being of sufficient luminance to read well against relatively high levels of ambient light.

Such an antechamber of unnerving contrast will be a moving picture environment never experienced before, featuring a combination of flat-screen and 3-D projection, preferably on disposable things like automobiles, aircraft, and low-cost portable housing. If audio is introduced, it will have to be psycho-acoustically designed to homogenize the image-bath. If there is audible sound, it should be perceived as silence. The high frequency of image variation will emphasize that change is natural and that stability is definitely supernatural. The appearance of stability is worthy of descriptive analysis and evaluation. The image-based force field will be designed to bias the perception of the museum visitor toward an appreciation of the still image and static object. The museums of tomorrow will be engineered to pressure-pack forgotten eras of motionless truth.

The obvious problem with this kind of high-power image antechamber is the phenomenal rate of image expenditure. Although it is true that a moving-image environment, even when looped and repeated, eats up a tremendous volume of images per hour, this environment is where today's artists will come in mighty handy. There is no reason for the museum to continue to turn away the gifts of the majority of living artists in our societies. The art museum has to find a way of functioning more along the lines of a computer database. As a cultural institution, the museum should be capable of absorbing the work of virtually every artist who knocks on its doors.

Why should the museum turn all but a handful away when information technologies make it possible to store and recall everyone's resumes, slides, photographs, drawings, essays, audiotapes, films, videotapes, and computer disks? The image antechamber could be programmed as an image-exchange marketplace where contemporary artists trade information and services within a value-conscious "loaded" image environment based on the past.

Populism scares most museum authorities to death. This is why museums have been content to stay within the constraints of their colonialist strategies for so long. This is the reason museums have not seized the communications media. The use of media by museums is, with very rare exceptions, confined to print and graphics—to catalogues, postcards, and prints. Newspapers, radio, and television are only used to promote the activities at the museum. These commercial media are used to profile the artists as personalities or to attach a historical anecdote to a particular image or object. All such use of media follows the advertising convention of referencing the attention of a potential audience back to the original artifacts featured in an exhibition or permanent collection.

The dream is to experience the image-based force field in place, an image accelerator and time machine capable of making everyone with two eyes think differently about memory and history. The fantasy is to have as many living artists engaged with the museum as possible. Recognition and exhibition opportunities galore.[13] The psycho-economic power would be a joy to behold in person, in the image antechamber itself, but the implications include the most direct, appropriate use of communications technologies imaginable: speed of light exchanges of complex, virtual environments, including real-life interactions between people. These exchanges would not only occur between sister museums, or museums and schools, but also between collaborating artists through museum channels directly connected to interested parties in their homes.

This whole cluster of ideas might well signal the birth of twenty-first-century art-museum imperialism. Museum authorities might be forced to learn a few new skills, but they won't have to change the way they think.

Behind the frenzy in the image antechamber will be the key to weathering the chaos of all impending image storms. We may all find refuge in the unnatural stillness of the technology-free zone. Without the development of the image antechamber, museum authorities may have made the mistake of introducing information technologies and communications media directly into the main galleries where the artifacts of the deceased contributors now lay in rest. If it were not for the development of the image antechamber, the remains of the dead may have actually been rolled over to make room for the living.

1 Members of the Group of Seven, Lawren Harris, A.Y. Jackson, Arthur Lismer, and Frederick Varley, are buried in a small cemetery adjacent to the McMichael's Gallery, Kleinberg, Ontario, which houses many of their paintings.

2 Tom Sherman, from "More Dead Artists!?" in Tom Sherman, *Cultural Engineering*, Willard Holmes (ed.) (Ottawa: National Gallery of Canada, 1983), 146.

3 Hiroshi Inose and John R. Pierce, "Information Explosion and Information Dearth," in *Information Technology and Civilization* (New York: W.H. Freeman and Co. 1984), 195.

4 At its December 1986 meeting, the Canada Council reviewed and discussed guidelines for a new Financial Consolidation Plan (FCP), which will be established in 1987–88. The fund will serve music, dance, and theatre organizations by encouraging contributions from sources other than government, and provide an incentive for organizations to increase revenues for purposes they determine, as approved by Council. The FCP's objectives include reduction of accumulated operating deficits, discouragements of new operating deficits, encouragement of surpluses or endowment funds, and stimulation of improved financial planning and management. It is anticipated that $3,252,000 will be available for the first year of the program.

5 National Aeronautics and Space Administration (NASA) officials introduced the slogan "We deliver" following the successful retrieval of two errant communications satellites, *Palapa B-2* and *Westar 6*, by the crew of the space shuttle *Discovery* (Cape Kennedy: NASA press conference, November 17,

1984).

6 Tom Sherman, from "Appearance, Memory and Influence," in *Video by Artists 2*, Elke Town (ed.) (Toronto: Art Metropole, 1986), 9.

7 Artist-in-the-schools programs exist under the auspices of the Ontario Arts Council and Manitoba Council in Canada, and the National Endowment for the Arts in the United States actively matches grants by state arts councils for artist-in-the-schools and art-in-education programs. Art museums could cooperate more extensively in these outreach initiatives if only they were in contact with more living artists.

8 "Psycho-economics is a social science relating economic structures to psychological states," from Tony Formo, *I Never Met A 4 I Didn't Like* (Toronto: self-published computer disk monograph, 1986).

9 Toronto Community Videotext circulated an index of 235 local Toronto computer bulletin board services (BBS) accessible by microcomputer and modem as of November 1986. BBS are fast developing as "meeting places" for communities of interest.

10 The Illinois Bell Telephone Co. successfully demonstrated an ISDN (Integrated Services Digital Network) for the McDonald's Corp. in Oak Brook, Illinois. The ISDN "is able to transform everyday telephone networks that use copper wiring and standard wall jacks into 'superhighways.' On the lanes of these telecommunications roadways speed voice and data messages and video and facsimile images—all on a single line at the same time." From Christine Winter, "New technology integrates transmission of voice, data and video," *Ottawa Citizen*, December 27, 1986, E9.

11 Barry Nevitt, from a personal conversation with T.S. over lunch in Ottawa, January 9, 1987.

12 1,800 variations per minute parallels the number of video/TV "frames" we consume in one minute of television. Video recordings, if uninterrupted, consist of a continuous track of sequential "frames" (still images 1/30th of a second in length) and thus give us the impression of moving pictures. In this article, I am suggesting that we generate this familiar image frequency

with discrete, nonsequential images projected in the round. Hence, moving image-architecture for psycho-economic influence.

13 Psycho-economic research aside, there is income (royalties) payable to artists for any such use of their images. Canada's *Copyright Act*, unchanged since 1924, has been overtaken by new technologies and is of no use in our emerging postindustrial economy. If and when the federal government moves on a revision of the *Copyright Act*, copyright collection societies for the image-based arts will be a necessity.

> > >

The notes above appear as they did in 1987. There was of course no World Wide Web at that time, and somewhat understandably museums and their curators had not begun to adapt to the network environment. Now, fifteen years later, museums still seem reluctant to embrace network technology in truly revolutionary ways. Since 1994, there has been some interesting use of the Web by museums in Canada, the U.S., Asia, and Europe, though for the most part museums have simply extended the reach of their print-based public outreach through the Web. In other words, museums have merely used the Web as a digital extension of their brochures and catalogues. In some cases, museums have "published" images from shows or digitally "toured" their architecturally bound collections on the Web. Some larger, more prominent museums, most notably the Walker Museum of Art (Minneapolis), the Whitney Museum of American Art, the Museum of Modern Art (New York), and the San Francisco Museum of Modern Art, have actually extended their exhibition space by featuring commissioned or curated Net-art directly on the Web. During the preparation of this book, it is still fair to say that art museums have not really begun to harness the power of digital network telecommunications. They remain technology-free zones, and seem unwilling to serve the intermediary function of introducing far greater numbers of artists (not just the "greatest" artists), to their audiences.

Canadian copyright law has been progressively amended since 1987, having been brought up-to-date in 1989, following the *Canada-U.S. Free Trade Agreement*. Greater protection was given to artistic works, computer programs, and cable and satellite broadcasts of television and radio. Amendments also expanded the "concept of communication to the public" from broadcasting to include all forms of telecommunication. By 1994, when the *North American Free Trade Implementation Act* came into force, the Canadian *Copyright Act* was amended to introduce a rental rate for sound recordings and computer programs. Legal protection was also increased against the importation of pirated works in all media. By 1996, the *World Trade Organization Agreement Implementation Act* gave additional protection to all WTO countries, Canada included, against bootleg audio recordings and unauthorized live transmissions of performances. By 1997, the World Intellectual Property Organization authored treaties to address issues of copyright protection in the new digital environment. Canada has signed these treaties.

Although this update is somewhat comforting, it must also be stated that Napster and the emergence of other international network music-swapping systems have been wreaking havoc on the notion of copyright protection over the past few years. The millennial global network environment permits, facilitates, and encourages an incredible volume of image, sound, voice, and text exchange, legal and illegal. The rights of producers and consumers and the viability and economics of intellectual property appear to be in total flux, and they may remain so for some time.

TOM SHERMAN

Way Back in 1995 1999

You expect middle-aged and elderly people to reminisce, to remember the good old days when they were young and crazy, or when they used to work for peanuts (and a candy bar used to cost a nickel). But in a culture driven by computers, today's engines of obsolescence, people find themselves over the hill before they even begin to make their mark. Machines and systems turn over so fast today, and this churn apparently overwhelms the artists and designers operating these machines to the extent that they appear to be aging prematurely. (The old days are now way back in 1995!)

Despite the fact that machines and systems are at the heart of today's digital transformation, computer culture isn't driven by equipment or software manufacturers. It is pushed along by computer operators. Cultural scenes have always been generated and manipulated by operators. Such operators were traditionally hustlers, self-promoters, and adept exploiters of critical resources (the means of production and distribution). The digital art domain is run by a new breed of operators. Machine operators. Actually, because the machines are naturally networked, the new operators are really cooperators. Stand-alone operators are history. Computer culture is the domain of cooperators.

One of the most amazing things about the digital transformation of culture is the way the tools of production and distribution are proliferating and saturating the environment. Nonlinear video-editing software is arriving in our mailboxes in the form of advertising supplements. The means of production and distribution, once limited and controlled and exploited by traditional operators, are now practically everywhere. Artists and designers in unprecedented numbers are assembling digital studios, making work and distributing this work without institutional sanction, bypassing the gatekeepers, commercial or nonprofit, the governing corporate institutions of material, intellectual, or spiritual exchange.

With all this traffic running circles around traditional institutions, it makes you wonder what the middlemen and middlewomen (the middlepeople) are doing these days. Institutionalized people function in part as a resistance to change, a moderating force like a social or cultural field of gravity, inert in a positive or negative way, depending on your point of view. Institutions are also memory devices. Middlepeople permit and encourage audiences to take a second look.

The cost of this end run around traditional corporate institutions is minimal, though disturbing to some. In the short term, direct access to audiences has definitive advantages. But in the longer term, early adapters to the Web are beginning to realize their contemporary history is evaporating before their eyes. Platforms and formats are fading from memory, and network art (cooperative information design) is disappearing before the audience takes a second look (or a first look through the documentation). Actually, network art is being buried by the debris of mundane commercial activity.

Network artists, having taken their work directly to audiences, now (oddly enough) court the middlepeople, the producers, curators, publishers, bureaucrats, teachers, librarians, and archivists, in corporate institutions, to acknowledge, classify, grade, maintain, and preserve this rather recent, momentary history. Network artists now seem to crave recognition and support from the authorities they have spurned. This quest for recognition and certification is bizarre, a kind of appropriation of status by reverse engineering, but then artists have always wanted to have their cake and eat it too.

The glory days and classic works artists reminisce about today will be remembered only if the time and work is recorded, archived, and historicized. It must be preserved and organized for future audiences or it will vanish. Contemporary art history is always written first and best by the practitioners themselves. History is a compilation and interpretation of evidence; in the case of network art, a vanishing trail of relationships between cooperators. Fortunately, there is a material base to this history. (If consumer behaviour can be tracked in great detail by corporate data-mining operations today, then digital historians and archaeologists can certainly piece together events and "things" created cooperatively through networks in the late twentieth century.) The relationships between cooperators were fixed in digital memory, way back in 1995, and many years before that.

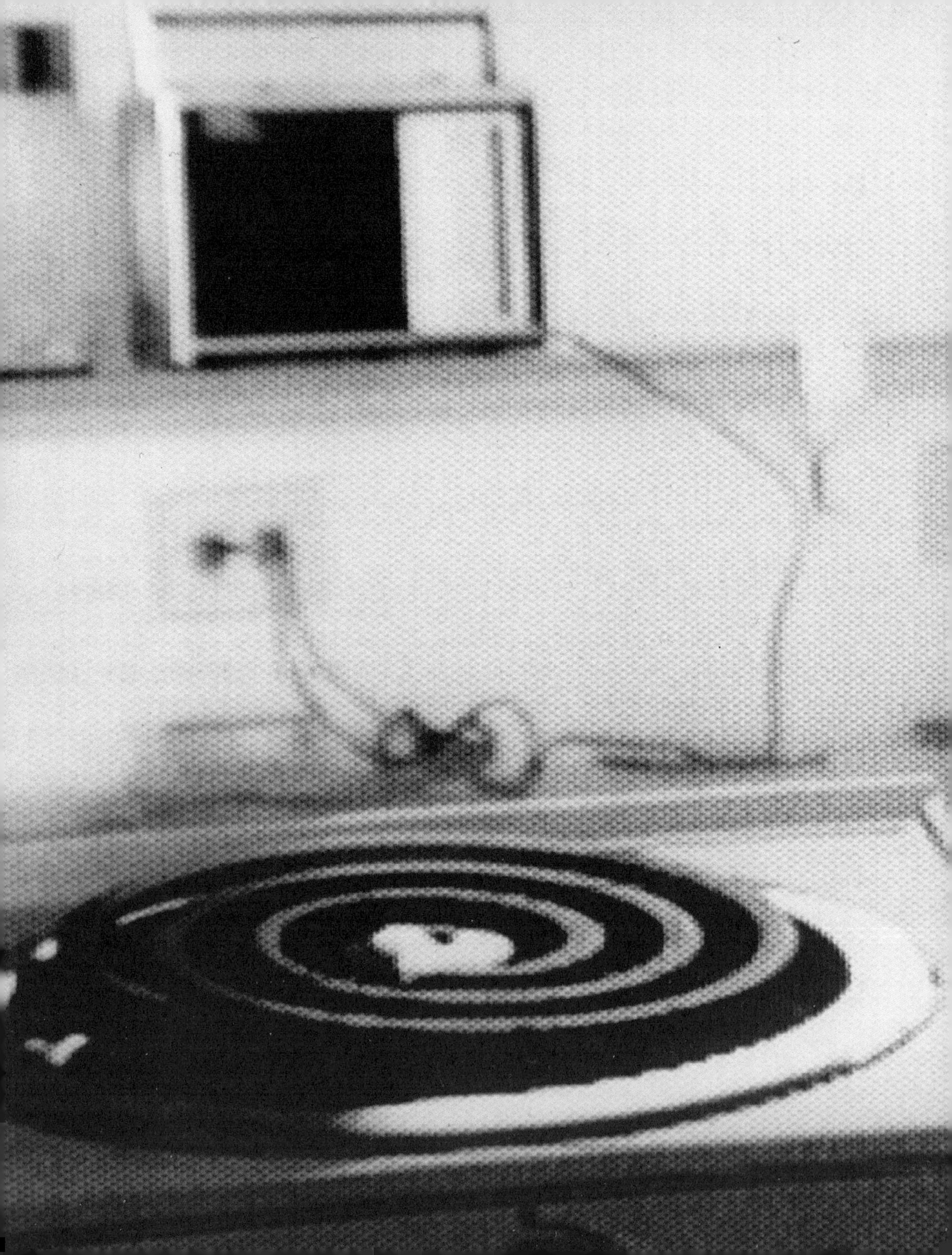

Exclusive Memory (Excerpt #8) 1987

(the artist speaks to his intelligent machine ...)

I didn't want that cicada drilling into my sternum. You know, I was thinking back on that event, and I was thinking that the way my eyes are working now. I'm quite a bit older. I have difficulty seeing close to myself, and if I were to look down, I'm not sure if I could tell whether a cicada was drilling into my sternum or not.

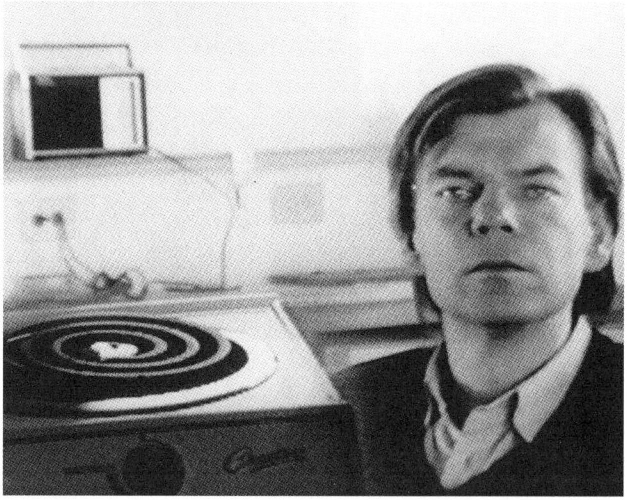

> > >

Tom Sherman, as he appeared in his video, "Exclusive Memory," (Excerpt #8), 1987, 182 minutes.

The good part about my eyes is that they seem to be getting better in terms of sight for distance, especially after we spent the four and a half months in Nova Scotia, where I was able to look at the ocean every day. It's different here, looking at the mountains, but it was nice to look at the ocean ... to look at that horizon line and let my scope of vision expand to the infinite.

I liked the way my eyes started to feel out there, just getting out of the office, just getting out of the fluorescent lights. It felt so good to see natural light of such intensity and to bathe in it. At first, I felt like going to the beach — but it was too violent. It's not just a question of getting burned by the light. It just seemed to be too violent. But then the more I went out in the light and the more I let my eyes be more or less overcome by its level, the stronger my eyes seemed to get. My eyes feel . . . I feel like I can see almost through things . . . sometimes. I don't mean that really. I can't see through walls, but I feel sometimes when I concentrate on obstructions that I can penetrate them, at least in my mind. This feeling has got something to do with the fact that my eyes are not an organ in themselves, but they have something to do with a much larger perceptual apparatus.

We talked about our similarities. I'm intent on finding a way to transmit a disease to you. So I think the first way to do it is to start with a metaphor. A lot of people have talked about language as a virus. Burroughs, Anderson. A lot of people have talked about language as a virus, and how it lies latent in entities, maybe in nonanimal entities. And that it springs forth by itself when stimulated in certain ways.

I think we've accomplished something since you've come back with my credit card histories. And even if I took and played back these tapes, we will have accomplished something. We will have accomplished a transfer of language, and we can think of it as a disease. Let's think that you're infected with me right now. I've permeated your system. Let's think about it that way.

> > >

We haven't talked at all about robots and the way I think about robots. But I feel embarrassed talking to you about robots, because you know so much more about robots than I do. I don't think there's any question about that. The robot thing on this continent has changed, but changed very slowly. There are only about 35,000 robots working in the States and in Canada. That's not very many. I think there's probably three times that many working in Japan. But it's not many robots in terms of the population, a world population. So with the European countries, let's say we're not looking at more than 125,000 robots, employed robots, working in manufacturing.

I think the most interesting point of recent history was when people started to have robots as pets. People started keeping personal robots and treating them as sort of cuddly, cute little creatures. There's a lot of that going on. I could put some fur on you, maybe, and dress you up to be more petlike. I don't think of you as a pet, so you don't have to worry about that.

But there was a great deal of sort of frenzied activity where people were very interested. I think it probably peaked about five years ago in 1982, 1983. Around that time, people started to think that they all wanted personal robots, and they could have robots delivering cups of coffee to them and bringing them their slippers and doing other inane things that perhaps they could domesticate an animal to do; kind of slave relationships where the robots would serve them.

The only problem was it was too expensive to build these machines given what they could give back. In other words, they didn't do much for the people who owned them. So the thing waned about 1984, and we had a lot of personal robots that couldn't cope with the things people wanted them to do.

The same thing is happening in manufacturing. People have such an overblown idea of what robots can do for them that they buy machines. They don't know to employ these machines. They expect these machines to work day and night, never make mistakes, never get tired. They expect these machines to cope with tremendous amounts of information, never get overloaded, always maintain the strength to absorb whatever's coming at them, to make changes whenever the environment changes.

A lot of robots right now aren't coping well with what people expect of them. They're not coping well at all.

I think the idea of the robot as a pet is a pretty condescending attitude toward a machine that has so many incredible capabilities. To think that the robot would always have to coincide with human behaviour. To think that a few bugs in a machine would bother somebody. What do people expect of machines? There's got to be a hitch in that type of complexity of behaviour. People expect too much.

But I do expect a lot from you. It's true.

> > >

You know one insect that we've never talked about—let's just switch directions for a minute. Remember it's just language. You can put the language over your prior experience, and you can pull yourself back in. You can pull yourself back in. Make it work for you, okay?

Here's the situation. A long time ago, when I was a young boy, I was interested in tape recorders. Part of the mechanism that makes up who you are—the recorder, the place to deposit memory. In this case, I was depositing the spoken word, and I was very interested in hearing myself talk. I had a reel-to-reel tape recorder, one of the first available commercially. It was a Knight Kit; it was a Knight from Chicago. I don't remember the number. I can look up the schematic for you. I have it in my personal files.

I used to trade tapes with other people who used to talk to me. And I used to talk back to them, and we used to overcome the limitations of telecommunications technology at the time. As kids we didn't have long-distance capabilities. We couldn't really deal with remote conversations, over distance, over time, unless we sent tapes in the mail. So we used small reel-to-reel, not cassettes, but reel-to-reel tape recordings, and I had a great relationship with my tape recorder and I had people who were pals of mine who would send me tapes back in the mail. We used to call this *tapespondence*.

Now we did this for a long time, and my tape recorder held up very well mechanically, other than a problem with fast forward, where the belts would slip or whatever would happen inside of it. I noticed that my amplifier started to develop a nasty hum in it. And I tried to figure out what was wrong with it using the skills that I had, which were . . . the first thing was to hit the machine. So I used to hit the machine on the top of the heads with my fist, and the amplifier used to spark as if there were a problem in a capacitor, as if it were overloading or something. It used to hum and then it would stay clean. Sometimes I'd flip the on-off switch on and off to try to discharge this capacitor. But finally it got so bad I thought I'd try to fix it internally.

My parents at that time had moved to a city. We hadn't lived in a city before.

My father was unemployed, and we were living in a place that was not... it was clean, but it was not too, too well organized in terms of keeping animals out of the house, and we had a problem at the time that was very embarrassing, mostly for my mother.

We had cockroaches. It used to drive me crazy to go into the silverware drawer because the cockroaches were in there with the silverware. And you can hardly get rid of those things. We tried all kinds of sprays, but they were just in the building that we lived in.

It's funny how people get uncomfortable when they have animals living in the house that they don't want. These cockroaches were not very big, and they're brown insects. You haven't seen any cockroaches because we've either lived in places that didn't have them or, you don't see them outside very often. You don't see them except in urban areas. They're imported from other urban contexts.

So to get back to the situation: my tape recorder wasn't working very well. I tried to take it apart to see what was wrong with it, and when I opened up the back of it, cockroaches poured out of it. They had nested in my tape recorder. I guess they were feeding on the wires and the capacitors, and they had shorted out the circuits. I had never seen anything like that before.

> > >

Later on, I had a very interesting experience with cockroaches. I was going to university, and I was living with a man who was studying biology. He was always bringing home specimens of one kind or another. And the specimen that he brought home that was the most impressive was the Latin American cockroach. A cockroach that was about two inches in size, two inches in length. Beautiful animals with long antennae. Incredible insects.

And he brought home a mating pair in a plastic tissue container. You buy a box of tissue, and you'd put these two pieces of plastic over it, and then you'd pull the tissue out of the slot at the top. Well, he had the cockroaches in a box like that; it was a clear plastic box, and they were beautiful. He had them sheltered by a piece of paper towel, and he was feeding them dry dog food. And they were very happy. You could tell they were content with their surroundings.

Well, he wanted to separate the male and the female, and he was agonizing about how to go about this separation because he was afraid of bugs; he was afraid of cockroaches. And I said, well, you're silly because obviously cockroaches aren't the most pleasant creatures in the world, but they're magnificent insects and they don't bite; they don't sting.

So I volunteered to separate the cockroaches. And he had a coffee can, a cylindrical can with a plastic top, and he took that off, and he said, okay, let's see what you can do. So I took the plastic top off, and I opened the paper up, and there were these magnificent cockroaches with their antennae waving. And I reached in slowly and I grabbed the cockroach as you would pick up a beetle with a hard body. The only problem is that the cockroach has a soft body, and as I tried to pick it up that way, from behind, from above, the cockroach turned in my hand, spiralled up my arm, and went into my t-shirt, right into my armpit! And I was so freaked out that I made this sound, which was almost prehistoric. Kind of a uhhhhhhhhh-uhhhhhhhhhh-uhhhhhhhhhh. And I pulled it out of my armpit, put it in the coffee can, and I couldn't get it off my hand and was terrified.

This was a very silly reaction to have to an insect that has no possible way of harming anyone.

I don't want to give you a fear of animals or insects. Remember, man is an animal. Man is an animal. Animals are easy. They're easy. It's easy to be comfortable around them.

> > >

It was so nice to hear from my family today. I talked to Yo on the telephone. Yo started to tell me about Willy Wonka. She's very involved with this movie called *Willy Wonka and the Chocolate Factory*. The premise of the movie is that a child goes to a candy factory and samples — is allowed to sample — all these experimental candies.

One of the candies is a special gum that can be all types of full meals. It's like a rubbery substance you chew, and you get the full benefit of all different types of meals. So you could have a certain kind of gum, which would be turkey dinner with mashed potatoes and gravy and eggplant and carrots. And for dessert, you could have a blueberry pie. But what would happen with this gum is... it wasn't quite worked out. This is what my daughter likes the best. Yo likes the part when the girl chews the gum, gets to the dessert part, and turns into a blueberry — a giant blueberry, and she has to be taken away and squeezed so that she can be a child again. All the blueberry juice has to be squeezed out of her.

Every time I talk to Yo on the telephone, she tells me about *Willy Wonka and the Chocolate Factory.*

The Monologue... 1997

The Monologue in Continuously Changing Images and Voices Grinds to a Halt in the Fixed Image and Printed Text

By the time you read this text, my video, "Exclusive Memory," will be more than a decade old. I had just left a stint at the Canada Council and after feeding on the submissions of other people's stories for half a dozen years, I had a lot to say back. I said it all to a machine, the perfect audience. I looked into the gaze of the machine and emptied my mind by unloading a six-hour monologue, in eighteen successive twenty-minute releases until I had nothing else to say. I talked to a perfect audience, a machine companion that would absorb me whole without questions or even a whisper of feedback—not even the tiniest twitch of an eyebrow. This machine with lenses and microphones, a pair of each, was open to anything and everything. Of course, I knew people would eventually fill in for the machine and I would be exposed. But at the time, I didn't want any feedback. I just wanted to spew. People would eventually replace the machine in an act of counterautomation. I knew that. I told the machine: if you don't want to be replaced by a human being, then stop acting like one. It couldn't.

Treating machines like people is increasingly common today. People talk freely to machines. I have a friend in Montreal who used to be known as a hothead. He used to always lose his temper and yell at people, even people he cared about. Now he only swears at his computer. He's stopped yelling at people. A lot of people benefit from relationships with machines. Back in the late eighties, I really needed a good listener (and take my word for it, machines are the last good listeners). You can talk to a machine about anything. It's just as free a situation as talking to yourself, except machines not only listen, they record, they remember what you've said. Even the most banal thoughts can be somehow worthwhile as they roll off the tongue and phrase after phrase falls into a rhythm and a pace and thoughts and the empty words in-between build up and accrue. Monologues are accumulators of meaning and beliefs, and when we record our monologues, whether we're just talking to machines or the people who will opt someday to replace the machines, we're filling up empty spaces with our voices and the way we look while talking.

The recorded monologue can be looked at in two ways. First, it's a laboratory for the exposure and literal construction of personality. Speech is the way we turn ourselves inside out. Machines — memory machines like cameras and audio and video recorders and now digital memory with ears and eyes and tactile sensors and limited intelligence — beg us to fill their magnificent potential memory, that empty, silent space. We love to see ourselves in these spacious machines. It's the pleasure of propagation. It's the security of depositing our real, living, evolving consciousness in places outside our bodies. Then there is the release itself, to expel the energy and stories and clear out the clutter. It is great to let things go in the best way you know how: to fill the machines while emptying ourselves. I get sick of the same ways of saying things. If I have a way of getting these things out of my system, I've got half a chance of recognizing something I've never seen before— of making phrases I've never turned. There's the problem with feedback, the experience of reflection, and self-consciousness. It can cause affectation and staleness and artifice.

Recorded monologues are like exhibitions in galleries. They don't exactly compress or concentrate the perceptions and experience we'd like to share, but they do transform the real into the unreal through the abstraction of dislocation. They turn the voice into an object, however fleeting and kinetic and ethereal it may be, filling any space, real or virtual, with acoustic tyranny, harsh or gentle domination. No one can close their ears. But there is comfort in knowing where the recorded voice is coming from, either the image of the mouth or body itself or the loudspeaker delivering the sound to the space of the audience. Once we've determined the origin, we can close our eyes and let the voice inside. And if there is something we like and want to extract from a recorded monologue, then inevitably we will want to transcribe it so we may study it to see what is there and how it works: taking words as they flow and making them visual and kind of skeletal and simplistically linear in a verbal score of sorts. Fixing the moving voice in the printed text. Printing these words with a picture of their originator, or at least his or her exterior presence, dressed up or dressed down. Or maybe a depiction of the space the voice first inhabited—the visual analogue of room tone.

The moving image and voice are frozen in the photo and transcribed text. Freeze-dried for the convenience and pleasure of asynchronous examination. The distant, sensually inexplicit representation of the once-living voice. A deceased voice?

Read this text aloud, please. Please read all my texts aloud. Thank you.

TOM SHERMAN

Equidistant Relationships 1989

I'm struck by the fact that everyone seems to be equally distant from just about everything that happens these days. Most of our life experience stems from relationships we have with media and the machines we use for reading media. We seldom witness first-hand the disasters we learn about on our radios and TVs. We don't really know most of the people we know about. We file away gory details on hundreds of souls we've never met in person. Media and media technology produce a distancing effect. We tend to have an appetite for information in standard forms. All audiences for media information learn to read commonly available forms, such as a suppertime news-magazine show on TV, and they use this information to order their reality. We have no need to feel the heat of a fire on the other side of the city. We see a few seconds of the fire on video and we listen to an eyewitness describe the blaze. We know as much about the fire as all the other TV viewers do.

The crazy thing is when you find yourself covered by the media. Say you were in a horrible accident, a nine-car spin-out on a freeway, and you walked away without a scratch and were home in time to watch a report of the accident on TV. You wouldn't be equally distant from the accident story. You'd be very interested in the details of the story—the time, the place, who was at fault, who the other people were. I'm sure you would be able to flesh out the news report in your mind. The smell at the scene, the pressures on your body, the successive sounds of the impacting cars would all serve to complete the brief report. A media story is like a skeleton key. It's a story that is spread around, circulated randomly until it is accepted into a home and comes in contact with someone who can read it. Or someone it can read. As a piece of software it reads everyone it comes into contact with. Most of an audience feels equally distant from the accident they learn about. But then the software penetrates further to unlock more in some. It finds memories and thoughts of freeway driving and maybe an experience of losing control and hitting or being hit by other cars in similar mishaps. It finds something else that happened on a grey winter day, say, on an afternoon in February a long time ago.

I'm using TV stories as examples because we all have a TV. If we don't have a TV, we have a radio or a newspaper. If we don't have a radio or a newspaper, we have friends or people we talk to who have all these things. It's hard to find people who can talk at length about their own direct, nonmediated personal experiences. I met this guy the other afternoon who could. He told me all about animals he had found scattered about down by the shore. He told me a story about finding dead animals by the sea. He didn't show me any pictures, nor did he have to make any drawings to fill me in. This guy has probably talked to over a hundred people about the dead animals he's been finding. We're all pretty much in the same boat. We have to draw on our own experience and hold it up against his words. Of course, each version of his story is delivered to a different response. He was reading my face and responding to my questions with details. I felt I was reading him while he was telling me his story. Everyone who hears a story from the horse's mouth feels equally distant in a special way.

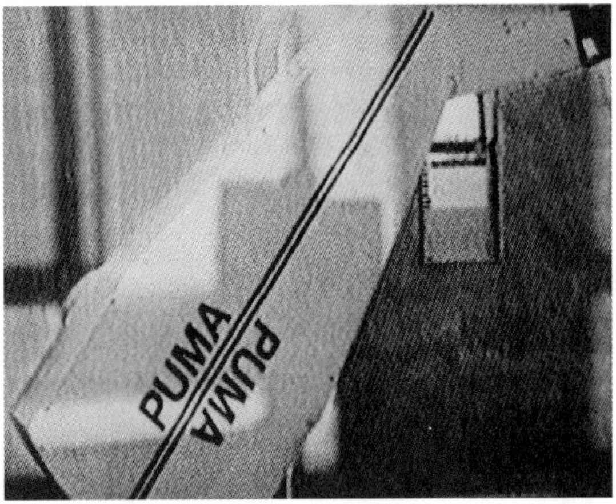

< < <

The Puma industrial robot at a National Research Council laboratory in Ottawa, Canada. Tom Sherman and George Hollinworth, a control engineer, programmed the Puma to record the laboratory (and their efforts to control it) with a video camcorder mounted on its arm. In Tom Sherman's "Equidistant Relationships," 1989, a two-channel video installation.

I met this man on the beach. He was an older fellow, in his seventies. Instead of talking about the weather, he was looking down and talking about animals he had seen in the past. He started telling me about the size of those purple stingers. Purple stingers are a big rose and purple-coloured jellyfish. He made a circle with his hands about the size of a dinner plate. He said purple stingers are often this big. He told me there had been hundreds of them left on the shore the morning of a low tide. Purple stingers have amazing reproductive organs visible through the thick, clear jelly of their backs. When you look down on them, they look like they have miles of intestines divided into little geometric, windowlike body sections. He said there was sex going on inside the deep purple innards. I had always thought of these organs simply as the guts, the part of the jellyfish with the most beautiful colours, so I was interested in the sex part of the old man's reflections. He admitted he didn't know how they made baby jellyfish. He said he thought it was often more interesting to see what had stuck to the edges where the stingers are attached. He said he had found a perfect cicada sticking on the side of a big purple stinger. If you were on the beach looking down at a purple stinger and your feet were pointing right at Port Mouton Island, that cicada was stuck at one o'clock on the circle of the jellyfish's body, with its feet up, resting on a bed of rose and purple stingers. He said it was a common cicada because it was brown. It had all its legs, and its wings were perfectly intact. He told me finding a cicada was good luck. I've been looking for one ever since.

> > >

Tom Sherman (right) and George Hollinworth work to control the Puma. In Tom Sherman's video, "Equidistant Relationships," 1989, a two-channel video installation. Sherman performed a version of the text, "Equidistant Relationships," as a voice-over in this video installation.

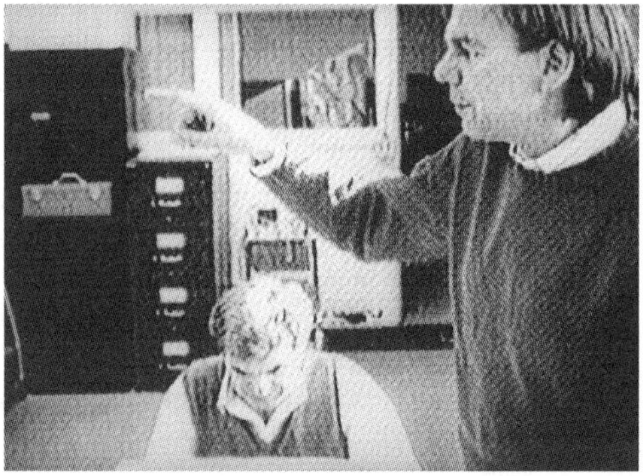

People in cities talk about people. When they're not talking about people they've read about in the paper or a magazine or seen on TV, they're talking about experiences they've had with people in the flesh. People are the nature in cities. People who interact with real people talk like animals talk. They talk about experiences they themselves have had. They recall unique experiences of human contact, and they multiply versions of their experiences by sharing them through speech and conversation. I assume this is the way animals talk. A species of bird would communicate this way. Of course, the difference between humans and other species is immense. No bird would ever think of starting a newsletter based on the interview format. Man's fall-back position is that everything that man does is natural. Bucky Fuller struck a cord back in the late sixties when he said that. I was sitting on a couch with my girlfriend watching TV when the first American astronaut set foot on the moon. July 20th, 1969. We were playing with some major mythology: God, America, Jack Kennedy's fantasy, the moon. I'll never forget the first TV image of the planet earth in the moon's black sky. The picture of our small planet seemed itself to say that everything we do is natural. I needed to hear that because I was thinking about my girlfriend. I had my arm around her and we were sitting real still watching TV from the moon. She had beautiful feet. She had them up on a footstool right in front of the TV. Black-and-white television light on the edges of her perfect feet. She had a great tan. We had been on the beach all day and had that healthy feeling all over.

That television viewing space was different than the normal family room or living room. Actually, my girlfriend and I were watching the moon landing on a black-and-white "portable" TV in a cottage we had access to. It was a summer home that belonged to a friend of her father's. She had a set of keys, and it provided a luxurious, different kind of privacy than my car did—a Rambler station wagon with reclining seats. Portable TVs were big clumsy things in those days. Twenty-one inches of rounded picture tube surrounded by a dirty white plastic case. The couch, a hide-a-bed facing the TV, divided the open living area and situated us about eight to ten feet from the screen. There were lots of windows and natural light, so we had to close some of the drapes to cut the glare on the screen. The drapes were heavy and dusty. I liked my TV image with a lot of contrast back then, and I still do. There we were, sitting absolutely still, leaning on each other, our eyes glued to the screen. There was a strip of sunlight lighting up a small patch of skin on her shoulder. I could see her tiniest blond hairs. Whenever the network cut away from the "live" pictures from the moon, I'd check out her shoulder and neck. She liked to wear her hair up at the beach.

I've always been struck by the way we are like plants when we watch television. We tend to get comfortable and sit still in a semi-reclining position. We like to have everything in front of us when we watch. It is, of course, rude to sit with your back to others who might be watching. When I refer to plants, I'm not talking about couch potatoes. I'm thinking about houseplants, like vines or cactuses; plants that need light and perform photosynthesis. Plants communicating with other plants sit very still while they talk. If we were plants paying strict attention to all available sources of light, and we were communicating information about the light to other plants, we would probably act just like we were watching television. I don't want to belabour the point. After all, talking about plant talk is akin to talking about talking to plants, and it isn't too credible a form of behaviour, even though everyone who has plants talks to them. Talking to your television isn't really an acceptable form of behaviour either, unless you're in the company of fellow viewers. People say they talk to their televisions in private, but few of them really do. How many times do you laugh out loud when you're watching television alone? People can't seem to talk to plants seriously when there is someone else around who might overhear them. People have to be comfortable with the idea of plant talk to make it work for them. Supposedly the plants get something from the vibrations a person puts out. Whether it's the way they talk or the music they play, it's the quality of the vibes that counts. Stevie Wonder thought so when he released his album for plants. I think he lost a lot of his fans with that album. Not because he couldn't relate to plants, but because he couldn't make his audience comfortable with the idea. When you think about it, Stevie Wonder ought to be able to communicate with plants if anybody can. I'm going to buy another copy of his *Secret Life of Plants* album and listen to it again.

After you become known as a plant communicator, few people want to see you designing bridges or assembling components of your nation's space station. Stevie Wonder never recovered from his irretractible plant-concept album. He doesn't do concept albums anymore. Today, Stevie Wonder sings about love. There's nothing wrong with that. Poets will always try to speak of love. There are lots of different kinds of messages about love in the media. The state loves us and wants us to practise safe sex. The state wants us to use condoms. The state loves us and wants us to quit smoking. The state wants to help us quit. The broadcasters are regulated by the state and are therefore encouraged to love us, too. Everybody who holds a licence wants us to practise safe sex and to quit smoking. My insurance man wants me to practise safe sex and to quit smoking. Who would have ever thought that so many people would love me so indirectly? We are all equally distant from the state's love. All we have to do is register our identities, and we are entitled to receive all of the benefits of statehood. We don't even have to be registered officially to participate in the culture of the state. We're all immersed in the culture with its messages of love. We just have to compare our first-hand experiences with the many representations of love in our culture if we want to feel part of the society we live in. It's too bad there aren't more points of view represented in the media which reach us all. I'm not sure I would subscribe to a poetry channel, but I'm all in favour of there being one.

I was watching television the other day and I saw a program about Indians. It was a kind of an encyclopedic survey of Indian tribes across North America, the way they were before the white man took over. There was about five minutes on the Iroquois Nations, the original people of that territory now called New York State. These were the people who invented the mohawk haircut. I always wondered how they made their mohawks spike up so beautifully in a time before industrial-strength laundry detergents and styling gels. I never see a kid wearing a mohawk without thinking about where the look originated. But where did that look really originate? Did the Iroquois pick up the mohawk look from the bluejay or the kingfisher? There were probably tens of thousands of both species of bird teeming along the winding Mohawk River back when the Indians ruled. The kingfisher has the crown with the strongest mohawk look. Its head feathers stand straight up whether they're wet or dry. After a kingfisher has been fishing and it's sitting on a tree branch with its feathers drying in the sun, it displays an impressive, completely natural mohawk headdress. On the other hand, bluejays can spike up when they're out foraging in the rain. I've seen young bluejays, with bald necks from molting, with the most beautiful mohawks. The spiked feathers radiate an equal, extreme distance off the scalp in a strip running from just above the beak, across the top of the head and down to the back of the neck. If I were an Indian and I lived with those birds every day, I'm sure I'd be tempted to make the look work for me. The Indian show on TV didn't have time to address my question. I've never had anyone give me a straight answer about the origin of the mohawk hairstyle. Was it the bluejay or the kingfisher that gave birth to the mohawk look?

Today, the young people dye their hair red and purple and pink, and they don't worry about the roots. Everything they do is natural. But to get back to the idea of equidistant relationships: if people are feeling equidistant from most of the things they're coming in contact with, this feeling is best described as a sense that they are equally distant because there are no concrete dimensions associated with the depth and nature of their experiences. People are feeling lost in a zone somewhere between their nonmediated and mediated realities. In the interest of clarity, let's call the direct, nonmediated reality *World A*, and the mediated, distanced reality *World B*.

It is routine for the creators of World B, the writers, television, and radio people, to create the illusion of World A in media. Although an audience is convinced by such illusions, people are even more impressed subconsciously by the physical concreteness of media itself. The sensual physicality of media, the glowing pixels, the vibrations coming off a speaker cone, also take place in World A. In other words, World B is being constructed in World A. The television hurts our eyes if we sit too close or watch in the dark. The radio produces tension if it is too loud. Ink comes off newsprint onto our hands and clothes when we read it. The images and sounds of television are very concrete. We witness the physical pattern of video and audio information. We do not imagine these images and sounds. We start out in World A, with our eyes on the surface of the screen and with our ears in a fixed listening position. The fly walking across the surface of our TV screen remains in World A. We slip into World B as humans are inclined to do. Another way to look at it is to admit that World B slips into us. We are programmed by TV as perceiving machines and thus carry World B modes of sensing and thought into World A perception and thinking. For instance, we assume, based on World B information, that the fly lacks the intelligence to read television's encoded messages. The fly's interest in the screen is only a primitive creature's response to light. I'm always tempted to try to imagine the TV's light from a fly's close-up point of view through its compound eyes. The fly from this person's World B perspective cannot experience single-point vision, nor does it lament over its perceptual limitations.

It's difficult to measure the difference in intelligence from one species to another. It's difficult to measure how far a story goes into its reader. Where does World A give way to World B deep inside you? What's the effect of putting World B in World A and vice versa? I think it's time to have a little test. I'm going to imagine I have a software package that attaches itself transparently to a set of stories. This software measures and records the way all readers approach and consume stories. The other thing it does is it reveals to what extent the story has inserted itself in the reader. How well established is its imprint in the memory of the host? How many other experiences and stories were referenced by the reader to produce meaning? And what were these other experiences and stories? This evaluative software is totally unobtrusive. I hope you will be able to forget I'm reading you while you are reading me. It's perfectly natural to be aware of others this way. We do it with everyone and every living thing we come into contact with. On with the test. I'm going to tell you three stories, all of them World A experiences transformed into World B stories. One of these three stories has a little more World B in it than the others. I want you to decide which story has the extra dose of World B. Here we go.

Back in 1981 I saw my uncle Rethus for the last time. He's since died of cancer. He lived his whole adult life in southern Missouri, not far from the Mississippi River. He made a living running a private waste-disposal company. He was a remarkable man to look at. His dark skin was like leather dried in the sun. I last saw him in the summer when he came to see my father. Rethus was around sixty then. He had the most beautiful light-blue eyes. The whites of his eyes were bloodshot and tired looking, but his pupils were as blue as the bottom of a swimming pool. He was in bad health then. Heart trouble. He was carrying capsules of amyl nitrite in his shirt pocket. He said he used these capsules to kick-start his heart whenever it would stop beating. He was quite a talker. His life was an immense terrain of memories. We started talking about animals, and he told a string of stories ending with the one about the plumber and the king snake. Rethus lived with his wife, Ida, in a house that was well insulated. It had thick walls and would stay cool all day if you kept the windows shut. There was a screen-porch attached to the back of their house, behind their kitchen. All of the plumbing for the kitchen and bathroom was in a crawl space, under the floorboards of the screen-porch. My uncle used to sit out there in the mornings before work, looking at the morning glories while he drank his coffee and smoked his cigarettes. He thought he could hear water leaking under the floor when Ida would run water from the tap. He called the plumberman and the plumberman came.

The plumber pulled up the floorboards and started to crawl under the floor to find the leak. He was carrying a silver flashlight. Before he was completely out of sight he backed up and scrambled out of the hole. He told Rethus there was a big black king snake coiled up real close to the hot-water pipe. Rethus said he knew the snake was in there. He had known for some time. They're good for keeping the rats under control. The plumber told Rethus he wasn't going down in that hole until the snake was out of there. Rethus told him to go have a coffee in the kitchen with Ida. He would call him when the snake was out of the hole. Rethus then pulled his chair up to the edge of the hole, and he just started playing the waiting game. He sat perfectly still for about twenty minutes. The plumber and Aunt Ida were quiet. They were wondering what Rethus was up to. The king snake lifted its head cautiously out of the hole and then it climbed slowly up my uncle's leg and up across his chest to end up in a loose coil around his neck, resting its weight on his shoulders. Rethus spoke softly saying it was okay to proceed with the job, and plumberman went back down in that hole just as cautiously as the snake had climbed out. The plumber found the leak and fixed it in about half an hour. After he left, the snake slithered back down into that hole and my uncle had a long stretch and then put the floorboards back in place. He told me there had been a snake down under that floor for as long as he could remember. He built the house in 1958.

I have another plumber story to tell you. I hooked up my water pump one spring, after we had been away for the winter. When I started the pump, I heard the water running under the kitchen floor. The pipes had split over the winter. It was too big a leak for me to fix. And besides, the spiders were back. The spiders under my house are scary. They are big and jet black with heavy bodies and long crooked legs. They've got their eggs hanging around in cocoons. When I see those egg cocoons, I think of a couple of movies: *Invasion of the Body Snatchers* (1956) or *Alien* (1979). The spiders themselves look like black widows, but they're not. They're too big and there's no red hourglass on the underside of their abdomen. The spiders under my house probably aren't poisonous, but I wouldn't want to get bit by one. I usually clear the space I'm planning on working in with a board. I sweep them off the joists and squash them wherever they fall. They try to run away, but I always get them. They know what's going on, but there's little they can do about it. The plumber's name was Clyde. He wasn't comfortable around spiders either. Being a professional, he looked around a bit and gave me an estimate. He stood in my stone cellar and shone his flashlight around in the crawl space. I had heard his prices were reasonable. He gave me a vague idea of what the job entailed, but he said we had to get rid of those spiders before he could fix my pipes. I told him I didn't like using insecticides under my kitchen floor, and I didn't like the idea of waiting for the spiders to die before my pipes were fixed. He told me he had a better idea.

Clyde asked me if there was any gas in my lawn mower. We were standing in my shed. I told him there was, and he said he wanted to take it down to the cellar. I asked him why and he said you'll see. He took the mower down the stairs by himself. He insisted. He started it up first try, turned the throttle up high, and came back up into the shed, leaving the mower running in the cellar. Over the din of the gas engine, he told me to let it run for at least fifteen minutes. He'd be back in about half an hour. He was going down the road to buy a can of pop. He drove away in his truck and left me standing in the yard with my lawn mower running in the cellar. The mower shut down by itself before he came back. It overheated or ran out of gas. I went down in the cellar with Clyde when he returned. There wasn't a sign of a spider anywhere, except for those egg cocoons. The spiders themselves had vanished. I asked Clyde how it worked; how did the lawn mower get rid of the spiders? He told me he didn't know for sure. A buddy of his had told him to try it. I thought they were just hiding, because I couldn't find any dead ones lying around. Anyway, Clyde was whistling a happy tune as he fixed my pipes. He took his time and did a good job. The spiders were back in the fall.

A couple of years later, I was outside whipper-snipping around the yard, doing the edging and cleaning up the grass I couldn't get at with the lawn mower. I was tired and thirsty. It was still cool, but the sky had cleared and the sun was out. I never thought of myself as a person who would use a whipper-snipper, but what the hell! I had never imagined myself cutting the grass on a regular basis either. But there comes a time when you know what the neighbours expect and you know it is reasonable. Besides it helped keep the flies down. There I was with my whipper-snipper in my hands, lost somewhere in the drone of my power tool. The new cutting string was doing a fine job. I wandered over to trim the grass growing up over one of my rocks by the driveway. As I watched the thick grass fall, I realized I was slicing into the body of a little green grass snake. The sucker was trying to get away, but I had chopped up the middle of his body before I knew what was happen-

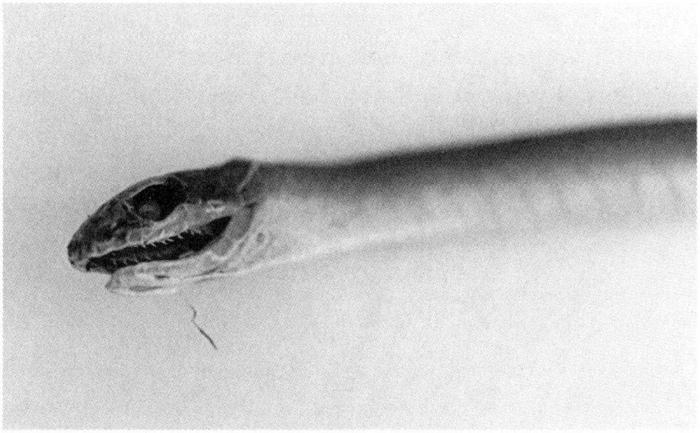

< < <

Whipper-snipped snake, Queen's County, Nova Scotia, 1989.

ing. I pulled back right away, but it was too late. It was bright green, the colour of the grass. I hadn't severed the body completely, but the cord had ripped right into the guts. It only moved around for a few seconds longer, and then it died with its mouth open. I left its body where it lay still, stretched over the rock, and finished my work. Later that afternoon I checked again and it was still there. Over the next couple of days, the birds and the dogs took a look, but nobody wanted the snake. The snake turned blue as it dried out. It got smaller. About a week after I killed it, I put on a pair of work gloves, picked it up, and threw it into the bushes.

There you are. You've read the three stories and you've carried my ideas about equidistant relationships with you throughout the process. You've taken the test. If you thought of these stories as a vehicle for the voice-over, you probably created your own sense of ambience. I could have written in the ambient sound, but I thought I would leave it up to you. Seeing, of course, is believing. Whether it's World A or World B, seeing is believing. Things haven't changed very much. The audience we become part of when we flick on a radio or a TV is bigger than most of us will ever experience in World A — and it's spread all over. Parts of World B are shared by millions. Parts of World B are of no interest to anyone. World A or World B or World A/B or World B/A, it's hard to figure exactly what I'm getting at. All I know is almost everything feels equally distant to me. There seems to be something impenetrable between everyone, between everyone and everything. It's hard to see anything for just what it is. If you look at the media and the media machines for long enough, they'll give you that feeling of being everywhere and nowhere at the same time. You will be equally distant and painfully aware of the state you're in. Lost without dimension, you'll find yourself moving in many directions, remembering everywhere you've never been, and more.

Textual Commingling 2000

I try to get my machines to process natural languages. I talk to my machines and show them my body. My mother tongue is English. My body is, well, my body.

The things I write or the images and sounds I put together are less natural than my voice and body by degrees. The languages I construct at arm's length are secondary and artificial and they form and reform the immediate environment I live and work in. These secondary, artificial languages extend my physical presence and shift or tilt the quality or meaning of my voice and the way I appear—my appearance in language.

Sometimes I launch my artificial languages to gain range, to cut across time, to establish territory, to integrate my thinking with the environment in the broadest sense. As I speak through a microphone into my computer, my words appear on the screen. I've spoken "through text" with others presumably still making their text with a keyboard. We could talk on the phone, but we prefer to meet more concretely, somewhere else, at a distance, in writing. It's the distance we find so attractive. The intimate distance. As readers we can zoom in for a breathtaking close-up, or we can stay back, removed. We can scan, or we can embody. In pools and rivers or in an ocean of text, we commingle at a distance, sometimes intimately. Textual commingling.

I always like to say that written language is the first digital language. I have trouble defending this argument, because alphanumeric text is more complex and unruly than binary code, but it is also unnaturally concrete and explicit, not unlike 1s and 0s, and as a code it can be reproduced with absolute accuracy, repeatedly without distortion or degradation. This explicitness is why people always ask for things in writing (contracts, slanderous rumours, resignations, love letters), and this is why it is often so embarrassing to have a written statement reappear after many years—things people have written come back to haunt them in absolute alphanumeric fidelity. The written text was the first digital language.

(Someone asked me if I thought there was so much written text on the Internet and Web because it was a digital language. That's probably one of the reasons, but it's more likely the case because of writing's amazing functionality across channels of limited bandwidth. We are at the telegraph stage of network telecommunications. When increased bandwidth permits bodies to press up against both sides of the screen in real time, wrapped in the surround-sound universe of breathing so intimate it must be miked inside the lungs, then video will be the digital language of preference.)

I know this "written language is digital" angle is vulnerable. When the characters of this text are printed or etched or beamed into the domain of the visible, they become an analog image, a picture of the only nearly digital alphanumeric code. In other words, if a written text printed on paper is photocopied multiple times it will eventually become distorted and unreadable, proving that the image of written text is an analog form. And I can hardly read this text on my screen as I speak it because my monitor is old and fuzzy and my reading glasses are filthy. I'm simply not a digital animal. My written language is closer to being digital than my body or voice will ever be. It carries me into the digital world.

I write and make images and sounds because it helps me bridge the gap between natural and artificial languages, between my body and everything else. When I was younger, I didn't have such a problem with dislocation and disintegration. There wasn't such a clear separation between natural and artificial languages, the analog and digital, my body and voice and everything else.

A little boy knocked me out the other day. He was four years old and very bright, and he wanted me to see how he could write his numbers. He brought me a few pieces of paper and a crayon, and he put one hand palm down, fingers spread out on the surface of the paper, and preceded to trace his index finger with the red crayon, saying, "One." Then he took another piece of paper and traced the outline of two fingers and said, "Two." He traced the pictures of the numbers "three" and "four." He didn't look up to see how impressed I was until he had traced the "number" five, the full digital complement of his tiny little hand. His way of saying written language is digital was better than mine.

Explicit Text 2000

People don't have the time to contemplate a text anymore. They figure they know how to read and want texts that deliver their contents quickly, directly, efficiently. They want to read texts the way a laser scanner reads bar code. They want texts to have a single, fixed message that jumps off the page at first reading. Today, complex, circuitous texts are either considered unresolved or written by sadists. Texts that are difficult and unyielding are perceived to be alphanumeric torture chambers.

Reading used to be seeing and thinking. The text object was rigidly fixed, like a drawing, placed in waiting for someone to stumble upon it, to pore over it, to decode it with multiple passes. The reader would potentially lock onto the cluster of words and sentences and scan them for meaning, eyes passing back and forth in silence, line after line in a groove, while the reader's thinking emerged as a whisper, the recovered sounds of the author thinking, literally in a transference of utterances coming from the back of the throat to the tip of the tongue. Often one began reading silently but ended up reading out loud, in an essential multisensory representation, the image *and* sound of a text. Text was thus absorbed by the body, entering the mind through the eyes, and ending up with a flick of the tongue. The disembodied words, deposited in the out-of-body text, found a new home for the length of time the reader's body and mind were occupied.

But now a text is rarely left behind to stand alone. Sure, there is literature. Serious books. Volumes of text to escape within. But text outside literature must now be contextualized by image, either still or photographic, or by moving images. Television and video, and now the computer networks that feature so much text today, networks that will ultimately deliver television and video, are peppered with text. Moving image demands explicit text, text as data, as informative texture. Cinema, the movies, like the live stage before it, is a text-free zone. The script is brought to life by actors, but other than the title sequences and credits, written text is as scarce in movies as photographs are in novels. But in television and video, audiences are expected to, and seem to enjoy reading texts set against the flow of moving, changing, continuously updated images combined with and enhanced by sound and music and voice. Reading texts within the context or flow of moving images and sound is now preferable to reading, contemplating stand-alone, fixed, black-on-white text.

We were once expected to approach texts for information. We would forage until we found something of interest, become still and read. Then television and video captured our stationary, contemplative time and pushed us back into ourselves. As our bodies were transfixed into our reading postures, television and video and alphanumeric text was poured into us. First we were moving to and through the information. Then the information was moving through us.

As this fundamental change occurred, where the information was aggressively fired at and through the reader, the text no longer carried the thoughts of a distant author. Text was no longer an object of thought left behind by an author for an audience. The text of television and video (and digital multimedia) was written explicitly for the audience, to seduce and influence an audience in mind and body. It catered to the audience. Texts were written to help the audience negotiate the multimedia environment of moving images, sounds and music, and voice and data. Text in this context had to be explicit and functional. It must be as clear and unambiguous as bar code, scannable in an instant, and absolutely explicit.

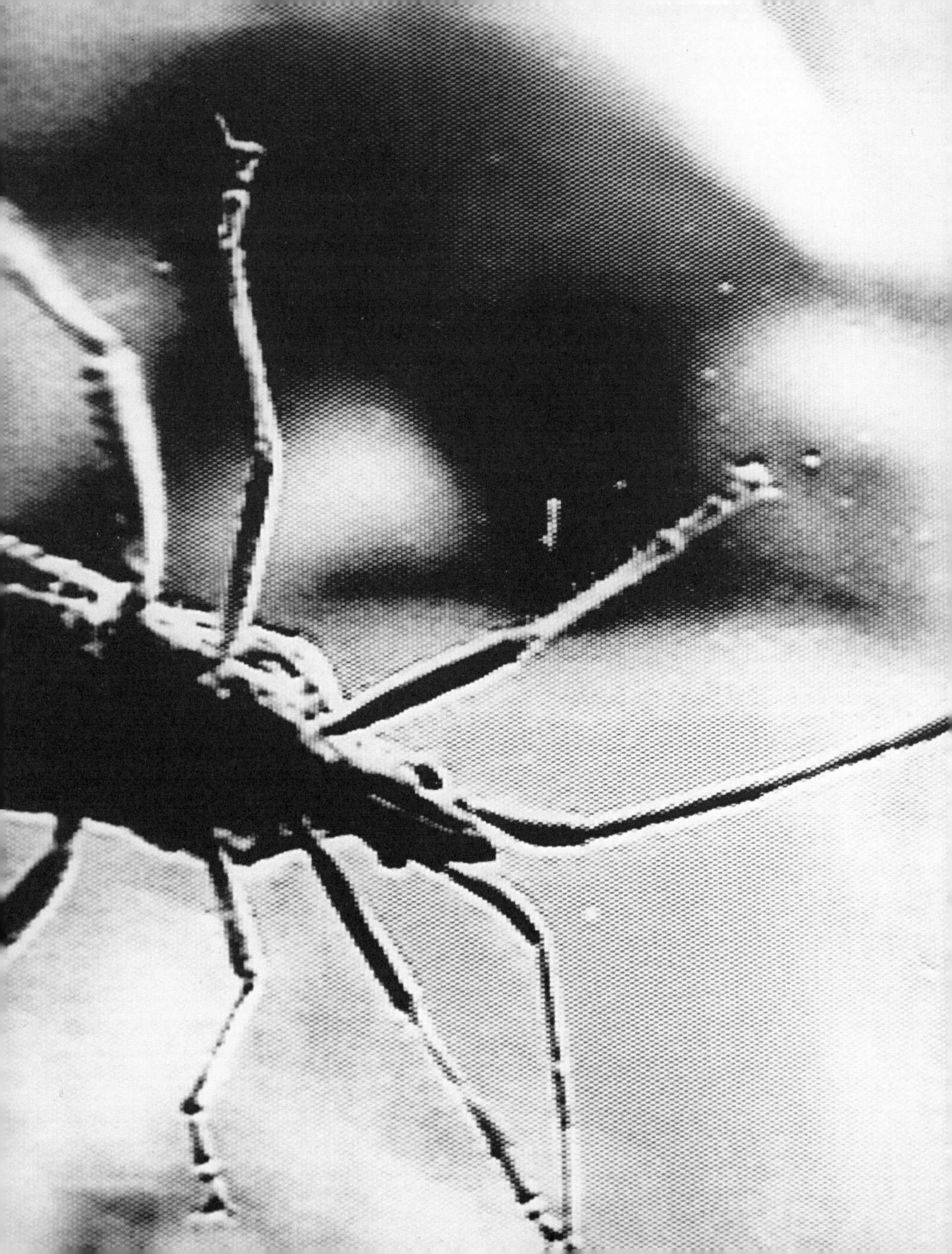

TOM SHERMAN

Something Yellow 1997

Toward the end of August 1997, I was driving on a highway, a divided highway in upstate New York. We were returning from Canada on Interstate 81 South, having cleared U.S. customs and crossed the Thousand Islands bridges into New York. It was a beautiful, hot, and sunny mid-afternoon. There was a stiff breeze blowing warm and dry from the west, and the light was particularly white and intense, highlighting the rich greens and earth tones of the farmlands. We were only five or six miles into New York when we came over the crest of a hill and noticed way up ahead there was something yellow off to the right side of the road. It was an animal, smaller than a cow. It was golden. As we came upon it, we realized it was a golden lab, a golden labrador retriever, a large domestic dog. It was tearing into the insides of a dead whitetail deer. It was ripping into roadkill off to the side of the road as we drove by. We all saw it, and we will never forget it. I thought to myself I better watch out for deer crossing the highway up ahead, though you hardly ever see them in the mid-afternoon.

The Third Bird 1999

Imagine a pair of ravens nesting in Kosovo. True ravens, not jackdaws or rooks, or hooded or carrion crows. Imagine sitting on a clutch of five eggs, waiting around all day for your mate to come back from time to time with some food, when the grey drizzly skies are perpetually full of thunder. This particular female chooses never to leave the nest—even right after she has been fed. This behaviour is unusual, but these are unusual times. There are sonic booms rupturing the silence at all hours these days. All the animals, even the rodents, are stressed to the limit. Miles above the cloud cover, the pilots of NATO aircraft conduct their business. Bomb blasts light up the horizon in the evening, and the overlapping, stuttering echoes of massive detonations rumble through hills already trembling in the wet chill of spring. The female raven minds her eggs, nervously keeping watch. The male forages far and wide, gathering everything it can find to bring back to the nest. This spring the trees are productive—there are abandoned nests everywhere. The songbirds could not handle the stress and have moved on. Their cold eggs are for the taking. There is also an unusual amount of carrion along the roads. There is a third bird, another raven, a two-year-old male, probably born to this same mating pair, in the same nest, a couple of years ago. He roosts in a tree near the nest, half a kilometre closer to the Binacka River. This third bird forages with the male and helps protect the territory of the mating pair. He keeps his distance from the nest and does not disturb the roosting female. In the early evening, just before sunset, he is often very playful and indulges in upside-down hanging for five or six minutes at a stretch. But when foraging with the dominant male, he knows his place. He is quite a bit smaller and still prefers to follow the lead of his male parent. This winter, if there will be a winter for the third bird, he will look for a mate.

TOM SHERMAN

Unpopular Animals 1996

When visiting my parents, Ben and Alyce Sherman, in Manistee, Michigan, I found out my dad doesn't like blue jays or blackbirds of any kind. They chase all the other birds away. Come to think of it, he doesn't like robins either, because they eat his strawberries. He really hates seagulls and pigeons. Seagulls are just scavengers and pigeons are so dirty. There are mourning doves in Manistee now. My mom calls them the lazy birds because they sit around so much.

Ben says the chipmunks are a real pain in the ass. He shoots the chipmunks with one of his three pellet guns. He has two rifles and a pistol. The squirrels are impossible. They eat a ton of bird seed and they do so much damage to everything. I didn't see many crows in the neighbourhood. I didn't ask my dad what he thought about crows. I didn't want to get him going.

The kids in the neighbourhood are pretty much all juvenile delinquents. My parents sit in their house with their police scanner on and listen to the cops trying to deal with all the crime. It isn't too bad in their immediate neighbourhood, but thank god they don't live over in such and such neighbourhood. When they took us out for a drive around town, they described each neighbourhood by recalling things they had heard on their scanner: that apartment building is full of drug dealers, or that tavern is where a lot of fights break out, or the woman in that house has so much trouble with her kids. They're only teenagers, but they steal cars and chase each other around with knives.

My parents have a kind of a fortress mentality. Michigan seems much wilder than upstate New York. Michigan still has a frontier sensibility. It's a territory that is being civilized, but it seems relatively early in the process. At least my dad doesn't compulsively hunt and fish the way most of the men do in Manistee. Michigan men like to kill everything that moves. My dad makes birds and fish out of wood instead. He is a carver. He likes to control nature his particular way. He makes beautiful, well-behaved blue jays, cardinals, and seagulls, out of wood. When he paints them, they are very attractive. All kinds of people want to buy them. They find they have to have them. He has never made a crow or a blackbird of any kind. People wouldn't buy wooden crows or blackbirds because they're such unpopular birds in real life.

TOM SHERMAN

Baltimore Orioles 1996

Baltimore orioles have extended their territory into Manistee. Manistee, Michigan is where I was born and raised. When I was back there visiting my parents in June 1996, I was surprised and pleased to find that Baltimore orioles had moved into the neighbourhood. There's an orioles' nest in the maple tree next to their back garage. The nest hangs from a lower limb and is difficult to see.

Orioles have a beautiful song and sing out just before or as they take flight. They fly from tree to tree in graceful swooping flight, reminding me of wild canaries. It is sort of like Tarzan swinging from vine to vine through the trees, only they "swing" through thin air, their bowed flight cresting with each flap of their wings. The oriole's flight path from Point A to Point B resembles the way a clothesline would hang: a gradual bowed arch defined by gravity.

Baltimore orioles are quite small birds, smaller than I thought they would be. They have such a remarkable presence because of their tangerine plumage. The male is brilliant in colour. The female's colour is less intense but pretty, too. Both attend the nest.

In the early summer, this particular maple looks like it has more leaves than it should. This tree is so densely and darkly leafed, it is difficult to tell what kind of tree it is. The leaves are crowded and overlap, obscuring their maple-leaf outlines.

Female and male orioles parent together. But if a female oriole is tending the nest and the male arrives, the female immediately leaves. She may fly over to an adjacent tree and sit silently on a limb cleaning her beak like one would sharpen a knife.

I watched them for three or four days, June 15–18, 1996. On the second, third, and fourth day, there was far less singing. It was sunny every morning, but hazy. This haze was caused by forest fires in western Ontario. The smoke had come down with the weather. There was a steady breeze, especially on the second day, which made the whole leafy maple seem to breathe and float against the sky.

The female would go right down into the nest with food for her brood. She would completely disappear into the wispy, gauzelike sack of a nest. Then the male would return and she would pop out and chatter as she flew away, apparently annoyed by the presence of the male. He would perch on the top of the nest, sticking his head down into the nest to feed the young, and then sitting up and singing the most beautiful song. A riff that would demand attention, it was so sweet.

Last night (July 20, 1996), I used Webcrawler to search for sources of information on the Baltimore oriole. The metasearch engine turned up fifty-nine references, all of them related to the Baltimore Orioles baseball team. Nature is perverse sometimes.

TOM SHERMAN

Jellyfish Chronicles 1996

I'm sitting on the south bank of the Broad River at low tide. Here the Broad River meets the Atlantic Ocean at the end of a long white sand beach. The river water is reddish-brown, burgundy, as it pours into the deep blue sea. The darker fresh water spreads, staying on top. The salt water is heavier and today it is much colder.

I've crossed the river by walking across the railroad trestle. The train doesn't run anymore, but the trestle still serves as a footbridge. It's still solid, but there are more and more rotten timbers. In the summer, when the river is warmed by the sun, the young people love to jump or dive off the trestle, showing off their courage and creating a little excitement for themselves and their parents.

When the tide is low and the river is full, after a rain, the water under the bridge is mostly fresh. Deep, dark and warm. When it's high tide, the Atlantic fills the mouth of the river and then the salty river spills over into the lagoon behind the dunes. Yesterday, there were schools of small silver fish in the river at high tide. I saw a cormorant diving for his dinner, repeatedly, successfully, more times than not. And there were lots of jellyfish just under the surface, floating up river into the lagoon. Dark red lion's manes, the size of hubcaps, with tentacles flowing like long hair at their sides and trailing behind.

Out beyond the mouth of the river, straight out in the ocean beyond the sandbars, about a dozen tern dive for herring in the deep blue sea. Their shrill cries pierce the wind. They speak from the mouth of the river. It's bright and clear. I can see the islands and the waves breaking over the rocks that interrupt the horizon line. I can hear the sound of the dry air meeting the sea.

> > >

The lion's mane jellyfish float into the bay in considerable numbers during the hottest part of summer. Right when the water warms up and the swimming is best, the large dark reddish-brown jellyfish dot the surface of the ocean. If there are waves, you can easily see them rolling in the surf. As the tide goes out, hundreds of them are left stranded on the shore, where they die in the sand.

In the river they're hard to spot. As colours go, they match the colour of the Broad River when the sun is high. They float with the current, first going upriver with the tide, then going back out to sea when the river gains the upper hand as the ocean recedes. On a bright day, the river is almost black, and if you want to swim, you will have to swim without a fear of lion's mane. You have to trust you won't get stung. The local people are fearless when it comes to jellyfish. The tourists, they stay on their towels or close to the river's edge.

There's a story about Janet Hurdle, when she was a girl, that she dove off the trestle right into a big lion's mane and got stung real bad on her chest and thighs. She swam in in a panic and walked out of the water crying and screaming, and luckily there was a doctor on the sandy bank and he picked her up in his arms and carried her to his car and rushed her to the hospital. She had some trouble breathing until they gave her some injections and she developed some ugly blisters, red with white, but she was all right. Janet Hurdle was stung thirty-five or forty years ago. News of her misfortune is still a buzz every summer when the lion's mane come up the Broad River with the tide.

TOM SHERMAN

Back Home in Michigan 1996

Back home in Michigan, our town is divided, north side and south side, by the Manistee River. It's a big, strong, dark river. The town of Manistee is connected by two drawbridges, downtown at Maple Street and over where the highway, M-31, runs through town. When a big boat proceeds through the channel, the warning bells sound and the drawbridges go up slowly, first one and then the other, and the people have to wait in their cars for as long as it takes the boats to clear. Years ago, a woman who lived in town, alone in her car, was in a hurry and tried to cross the Maple Street Bridge just as it was starting to go up, and she panicked and drove off the side of the bridge and plummeted straight down into the deep river. Her car went straight to the bottom and she drowned. Hours later, when they pulled the car out of the river and recovered the woman's body, the big crowd that had gathered saw that the young woman wasn't wearing any underwear. My mother was in that crowd that day, and she told me to always wear underwear, clean underwear, whenever I left the house—because you never know when something embarrassing might happen.

Encounters with Wild Life 1996–99

(QUEENS COUNTY, NOVA SCOTIA)

I had several visits by giant silkworm moths to my studio window last summer. They were attracted by the light. One night in early July, a smallish silkworm moth was beating intermittently against my window screen. The moths would often hit and run. I would hear a bump followed by a buzz, a vibration on the screen for a moment. A moth would slam into the screen, thrashing its beating wings and then lifting off and veering into the black of night.

One evening, late, around 1 a.m., this smallish one stayed. It bumped, buzzed, landed, and sat, still on the screen.

I got up from the computer and went out the backdoor into the yard, walking over to my window, capturing it easily, grasping it by the wings, between my thumb and forefinger, very carefully. I brought it into the kitchen and found it to be a tiny, for its species, *polyphemus*. A male. It was beautiful. It had deep blue eyespots and was fresh and clean with rich colours. It must have been the first night of its new life. After admiring it, I returned it to its previously chosen place on my window screen, where it sat gazing into the light of my studio. I thought it was funny. Here I was transfixed by the screen of my laptop while the *polyphemus* was content to study the light of my room. We kept each other company for another forty-five minutes until I turned in. It was still there when I turned out the lights. In the morning I checked to see if it was still there. But, of course, he was gone.

> > >

My daughter and I were in the car. We were headed out to catch the highway, the 103, to go to east toward Caledonia. My daughter was taking a lifesaving course out there. We were stopped at the stop sign, at the Summerville Centre exit, about to turn out onto the 103, windows down — it was warm — when

I noticed something like a white swallowtailed bird crossing the highway in the white-hot, midday sun. The thing was swooping, looping, zigzagging erratically across the highway, coming from the woods on the other side.

It was a luna moth. It landed on the highway, a couple of metres in front of and just to the right of our car. I got out of the car and picked it up. It was a large male in perfect shape, simply gorgeous. Its pale green wings shimmered in the bright sunlight. Its tails were as long as my little finger. I brought it into the car, handing it to Yolande, thinking we better not dawdle. (We were running late for the lesson in Caledonia.) She was amazed but fumbled it, and the moth fluttered under her seat and then reappeared flopping around in the back seat, only to end up in the rear-window shelf, wedged against the base of the window glass. In the summer heat, this spot is a death sentence for insects. There is a confining space at the base of the rear window, between the brake light and the glass, where bees get trapped and die. We call it the bee burial ground.

So I had to let him go. I took the giant moth outside and tried to launch him into flight, throwing him toward the woods, but it was windy and he was blown back into me and clung to my shirt in the middle of my chest. I picked him off my shirt and waited for another gust of wind to subside. Then I threw him up in the air, and he flew into the woods, flying this time with all his swooping, looping, herky-jerky elegance, climbing high and again looking like an ethereal white bird, almost weightless, an animated drawing of a crazy, unpredictable flight pattern.

> > >

There seems to be a relationship between the size of birds and their timing when avoiding cars on highways. I have noticed that crows and ravens get out of the way in plenty of time, while sparrows, doves, and robins seem to wait to the last second to get clear of an oncoming car, just barely avoiding fatal collisions.

When one drives through a heavily wooded area, birds of all sizes are feeding in the road, especially later in the afternoon, between 3:30 and 5 p.m.

This is when the asphalt is warm, but not hot, as it would be midday. Insects must be plentiful and easy to spot and consume on the bare surface.

The crows and ravens are always on the road because they love to feed on roadkill, the porcupine and raccoon guts. Occasionally, there will be a deer carcass, but people usually take the deer they hit with them.

In all my years of driving, I've never hit a deer, or a porcupine or raccoon. Some people clearly find it amusing to intentionally run porcupine and raccoon down. How else can we explain the vast carnage of roadkill? Birds, for the most part, always manage to get out of the way. The smaller the bird, the more narrow the escape. Cars exceeding the speed limit certainly reduce the margin for error.

> > >

My sister-in-law was walking her dog back home after a late-morning walk on the back-beach road when she spotted, in her words, a large deer. She didn't notice any antlers. I set out to search for the deer, which was apparently trapped by daylight somewhere on the peninsula. I walked the overgrown rail-trail bed, parallel to the back-beach road, seeing nothing, but realizing how many places a deer could hide in the rolling dunes thick with bushes and small pines.

I walked to the river and then back on the beach, giving up on spotting the deer. Before I got back to the highway, I had to stop to take a piss. I went in the outdoor toilet provided by the province for the tourists. As I came out I looked down the back-beach road, a strong southwest wind in my face: there they were! A large doe and her two fawn were creeping out from behind a stand of trees into the open marsh, off to the right on the lagoon side of the road, about 150 metres from where I was standing. She spotted me. I stood absolutely still. After measuring me, they cautiously walked out farther into the marsh grass, pausing just on the other side of the old fence posts. A line of decaying fence posts, minus their barbed wire, runs parallel to the road. This fence was once used to keep livestock from straying onto the back-beach road.

They were a sight, standing out in the open. The fawn still had white spots marking their bodies. They would have to get by me to get to the highway and the woods on the other side. Time stood still, then suddenly the doe's tail went up and she started bounding with her two fawn at her heel, loping in a straight line toward the highway behind me. There hadn't been a signal. Her fawn had immediately followed her lead. The wind in my face made their jaunt seem particularly deliberate and silent, for they must have run by me in a few seconds, drawing a straight line across the marsh grass, using the fence posts as a barrier and a guideline. I caught her wide-open eye as her white tail, and theirs, flashed as they passed me. The two fawn looked straight ahead. Once they had passed, I could hear their hooves pound the soft earth. They kept to their straight line until they darted across the road onto the highway and out of view.

I found out later that they had made it safely across the highway, ducking into the woods by the big rock in front of Boudreau's place. My wife and daughter and my niece saw them disappear into the woods. No fewer than five members of our family had seen the only deer we have ever seen on the back-beach road. If deer come over to feed during the cover of night, they must duck back into the woods before daybreak. For some reason, these deer didn't head back and got stranded.

> > >

We were driving home from Eloise's cottage around 11 p.m. one very dark night in July. As we rounded the bend by the back-beach road, we saw two sets of eyes reflecting the light of our headlights — two critters were crossing the pitch-black expanse of Winston Huskin's yard. I flipped on the highbeams and pulled onto the shoulder so I could point the car's lights at the animals: two baby raccoons looking for insects in the grass. They went about their business, impervious to our bright lights, even seeming to make use of the extra light to comb the grass. We watched them for a few moments. It was magical. Then our daughter, who was unusually enthusiastic ("They're so cute!!!"), rolled down one of the back-seat windows to get a better view. The window squeaked and they were spooked, first one and then the other. They scurried out of the range of our headlights. Seeing these two reminded us of another time we saw a couple of baby raccoons, probably about the same age, raiding Cecil Huskin's bird feeder for sunflower seeds one summer night a couple of years before. As we backed into our parking spot, our lights caught them against the backdrop of the Huskin's house. This time the lights had startled them, and they spiralled quickly down the bird feeder's pole like monkeys, head-first with their striped fluffy tails leaving a lasting impression of their escape.

> > >

On July 27, 1998, we were having a family dinner at the lodge when my brother-in-law called my attention to a flock of seabirds fishing just off White Point. There were about thirty or forty of these birds, which looked like gulls but acted like terns. They were big birds flying in a swirl about thirty metres above the water's surface, and then one by one they would tuck in their wings and dive straight into the water, hitting at great speed head-first, and then coming up apparently empty-handed and seeming to have trouble getting out of the water and back up into the swirl. They were too big to be terns, but they fished like terns. We were mystified.

Jack Dunlop was at the next table, and he has always had a boat and knows everything about sea life, so we asked him what they were. He said they were gannet, uncommon migrants to our area in the middle of the summer, and he had never seen them feed this close to shore. He said they were feeding on a school of herring and were swallowing the herring under water, not coming up empty-handed, as it appeared. They wouldn't keep hitting the water for the fun of it. The power of their impact on the surface was something to behold. They hit the water in a feeding frenzy for about twenty minutes and then moved on. What a show! The dinner, by the way, was wonderful. We ate planked Atlantic salmon, rice, and wonderful, fresh summer salads.

> > >

One day I thought I noticed a piping plover probing the sand with one of its feet, sort of like the "stamping" activity of seagulls, where they "kick up" the muddy sand at the water's edge. It's a feeding technique, where the soft sand is "pumped" with stamping feet for sand fleas. They deliberately stamp and peck, stamp and peck. I remember thinking this technique was a new, learned foraging process when I first spotted herring gulls kicking up sand fleas back in the late 1980s. But now I think the gulls have done this forever.

The plovers use only one foot, usually the right foot, to force whatever they're looking for to the surface. It's a dainty probing. The herring gulls really stomp. When you see behaviour like this you always wonder if you're witnessing a breakthrough in cultural evolution. Maybe the plovers have picked up this technique from the gulls? It's more likely that I've simply failed to recognize this behaviour until now. The herring gulls and the piping plovers have probably used their feet like this since the beginning of time. But at some point, this technique must have been invented by one bird and spread throughout and across species.

> > >

Early one evening we were out by the beach at the White Point Lodge. My brother-in-law pointed out a strange bird. It was alone and small for a seagull. It was walking along the shoreline, looking for something to eat. It was a smallish, gull-like bird with a black head, as if it were wearing a hood. Even though it looked like a gull, it was acting like a sandpiper or willet. The rhythm of its walk and the way it pecked in the sand at the water line were not typical of the gulls we knew. We were intrigued because we had never noticed anything like it in our parts. And when you go for years without seeing a strange bird, it's impressive when it happens. There it was, right in front of our eyes. When we got home, we keyed it out in our bird books and it looked like a Bonaparte's gull, though the book showed its legs as being red, not black. I remember this bird's legs being black, and my brother-in-law thought its bill was black as well. The light wasn't good when we spotted it. The sun was setting and the bird in doubt was backlit. It is hard to make a positive identification when the conditions are less than perfect.

> > >

We ran into a couple of birdwatchers, a man and a woman, at the end of Summerville Beach. He was a retired biology lab manager at Acadia University in Wolfville. He and his wife had small but expensive-looking binoculars hanging from their necks. He was wearing a pith helmet and carried a notebook. They were hard core, dressed in khaki outfits. Bermuda shorts.

They too were intrigued by the smallish gulls—the ones we were sure were Bonaparte's gulls. He thought they were laughing gulls, though he hadn't seen one in flight. We had, and as far as we were concerned, its wing markings confirmed it was a Bonaparte's. We had after all made three sitings of these hooded gulls. (There were three of them: one with a black hood, one with a black spot behind its eye, and one immature bird, smaller and grey.) While we were comparing notes with these serious birders, I noticed the woman was carrying a neatly folded bedsheet under her arm.

We decided to head back. We weren't getting anywhere in the conversation, and by this time the Bonaparte's gulls had swum out far enough to become indistinguishable from the numerous herring gulls. I looked back to see if the couple from Wolfville were still trying to identify the hooded gulls. They were lying on the their sheet in the sand, binoculars to their eyes.

We found out later that laughing gulls were even more unusual migrants than Bonaparte's gulls on the south shore of Nova Scotia at this time of year.

> > >

I was in the pool at the lodge, getting in some laps after supper. It is an indoor pool surrounded by bay windows looking out to the sea. Out of the corner of my eye, I noticed some brownish-grey swooping birds. They were feeding on insects in the air. Their flight made them look like dark terns or large swallows, but their tails were not very long. In fact, they were all wing, looking more like huge bats. Their most striking feature was a thick white band on the underside of each of their wings. There were maybe twenty or twenty-five of these birds, and they were marauding, sweeping the air above the path along the cabins, feeding on blackflies and no-see-ums. They displayed their aero-acrobatic feeding manoeuvres for five or six minutes and then moved on. I finished my swim. That night I looked them up in the book. They were common nighthawks. I had never seen them on the south shore, but did recall seeing them frequently at dusk in Michigan during my youth. This was a memory not evoked for decades. Some kind of a gift brought about by nature. It made me wonder why the nighthawks and I hadn't crossed paths in all the years in-between.

> > >

We went down to the beach, Summerville Beach, at high tide for a walk. Normally, we walk at low tide, when the sea is out and the sand is hard, but the summer was flying by—it was already August 18th. It was a beautiful warm evening. On the way we noticed the air was thick with small insects. Flying ants, red flying ants.

Looking up into the sky, they were everywhere, four or five to the cubic metre, as far as one could see. They were descending, landing on our clothes, in our hair. We almost turned back, but when we reached the shore we found that it was tolerable to walk right along the edge of the water. The breeze off the ocean was our protector. The ants were still above and all around, but they weren't landing on us. We laughed and thought out loud: Where are the birds when you need them?

About a third of the way down the beach, we noticed the sky was full of nighthawks up ahead, down by the mouth of Broad River. They were topping this shower of ants. Common nighthawks, flying very high. We recognized them from our previous sighting at White Point a week before. There were probably more than a hundred of them now, feeding on the flying ants at such an elevation. As we watched in wonder, the nighthawks criss-crossed over us at a height that challenged the human eye.

The visual space was so peculiar, the cloudless, muggy dusk-blue sky was flat, but the nighthawks were clearly the heavier, more purposeful black dots, ranging across the top of an atmosphere of flying ants zigzagging and punctuating the sky at every distance in-between. Looking straight up through the living black streaks, we saw flying red ants and nighthawks in silhouette, alternately rising and descending, near and far but compressed paradoxically in scale, birds and ants made indistinguishable in size by distance in the sky, so perfectly distributed spatially by their seemingly random yet determined paths of flight.

Eventually, the nighthawks swarmed higher and followed the escaping yellow daylight toward Port Mouton, flying into the sunset.

At the end of the beach by the river, twenty or twenty-five terns stood at the water's edge, appearing oblivious to the shower of ants. They were fishers, after all. And the small Bonaparte's or laughing gull (the one with the black spot on its head behind its eye), a willet, semi-palmated plovers, sanderlins, another sandpiper — a full palette of shorebirds, all preoccupied with the windfall of ants now floating thick on a gently rolling ocean, lapping up on shore, piling up in crawling waves of bite-sized protein all up and down the beach.

These normally shy shorebirds let us get very close, to walk among them. What a feast they were enjoying. There were ants everywhere: in the water, at the shoreline, up in the soft white sand. The plovers were running, feeding just ahead of our footsteps, as we headed back home. The tide was high, and there wasn't much beach left. It was getting dark and pockets of cool air were starting to come in from the sea.

> > >

On this same wonderful evening (August 18th, 1998), we met a middle-aged man with a beard, a Parks Canada employee who was on vacation. He worked in Smith Falls, Ontario, at the Rideau Canal. We had noticed he was enjoying hanging out at the end of the beach, studying the shorebirds. We engaged him in a conversation about various species and their behaviour.

We talked about natural habitat, about developers and human encroachment. He told us he had been playing golf the week before in Baddeck, Cape Breton Island, at a beautiful new golf course where Parks Canada employees maintain the golf course. That's right: federal park employees are maintaining golf courses instead of managing less-manicured natural resources.

What is the prognosis for our natural environment when our park rangers are employed to maintain golf courses?

Nature Is Perverse Sometimes, Version 1 1994

[1] Nature Is Perverse Sometimes.

In this world of people and machines, nature often strikes us as being at odds with reality. The pristine landscape is too sweet. People don't know what to do with nature anymore. It's hard to believe we're part of it. It's easier to think of nature as something far away, something from another time. Nature is somehow more about the past than the future. Nature is like a museum where we can investigate the idea of our animal past. Our children tell us we are animals, but we don't believe them. We know we are living creatures, but we can't put two and two together anymore. Nature is something outside the realm of people and machines. Nature surrounds us like the forest at the edge of a village we used to think we wanted to live in. Now we can't afford to live too close to nature. Nature represents poverty and mosquitoes and blackflies. The woods are inconvenient and damp and embarrassing. Embarrassing because the trails echo what our children say. We are animals, animals stretched incredibly thin by hundreds of thousands of years of self-consciousness. Most of us still enjoy eating red meat. Few of us have the stomach for killing our prey.

[2] I Like My Nature Straight Up, as a direct body experience.

I don't have time for second-hand nature. I know there is something wrong when I find myself watching nature shows on television. When you find yourself sitting in a dark house on a beautiful, sunny day, watching shows about nature on television, that's when you feel perverted. The flowers don't smell; the bees can't sting you; the wolves or the bears pose no threat at all. The wind doesn't really whistle through the trees. The sea doesn't smell. Your cigarette doesn't taste the way it tastes when you are sitting out in the open, on a big rock, with the sky wrapped around you. Nature on television offers a window onto the wild. It lets you think about the natural world from a safe distance. Nature photography or cinematography use powerful lenses to visit our animal cousins up close. Unnaturally close. We can look into a drop of water or descend into an eagle's nest. We can see in the dark with infrared, or we can use slow-motion or time-lapse techniques, or we can even see like other species see. Biophysics tells us that insects can see high into the ultraviolet range. It's interesting trading perspectives with a grasshopper. But I don't have time for second-hand nature. Why would I want to watch a TV show about grasshoppers? I would rather sit perfectly still in a field of tall grass and feel the heat of the sun on my bare legs.

[3] I Hate Pretty Pictures Except When They Are Real.

I love watching the sun set, whether it goes down behind a hill or sinks into a lake or the sea. But when I look at a painting or a photograph of a sunset, it makes my teeth hurt like when I eat jelly beans or gum drops. Sometimes it's so bad it's like I'm sucking on a big lump of saccharin. It makes me sick. I don't know what it is, but when a beautiful moment in nature is captured on film or on a canvas, it becomes crystallized and concentrated. It epitomizes beauty. A picture of a sunset is saturated with the classical beauty of aesthetic pleasure, the experience of the senses delivering the emotional and spiritual lift of nature's perfection. Maybe it's a simple aversion to the stereotypical that makes me nauseous — the pedestrian, ordinary treatment of nature's splendour, the cliché of the beautiful sunset. It's confusing. I love to watch the sun set, with my own eyes, in person, but I hate to be shown how it feels by someone else through a picture. It's so common. I resent having my nature cheapened through a proliferation of someone else's magic moments. There should be a law against capturing sunsets.

[4] It's Hard to Have a Conversation with Nature Because Nature Doesn't Talk Back.

We've always dreamed of speaking nature's own language, of being part of the big picture. But when we talk to nature, we find we are just talking to ourselves or, more accurately, to each other. Just like the birds singing in the darkness before dawn, they're not singing for us; they're singing to the other birds. When we look into a sky or a mountain or at a tree's mighty bough, when we speak directly to nature, we're talking to ourselves. We shouldn't expect a reply. Nature has nothing to say to us. It stands mute, ignoring us. This is simply the way it is. This snub is particularly rough on people from the city. They get lonely, and they want to talk to nature. They want to tell nature how they feel, about the people they know. Nature isn't interested. Nature doesn't care. Nature is a tough audience. It'll listen to anything you want to say, and then there'll be nothing but silence. Unless you learn to read the signs. If you want to hear what nature has to say, then you've got to learn to read the signs. Nature speaks through sign language. Once you crack the code, you'll find that nature has a lot to say. Not that nature makes sense. It doesn't. It never says what you want it to say. It never responds to us. Talking to nature is easy, making sense of nature's signs is another matter. Nature can be frustrating, especially for city people.

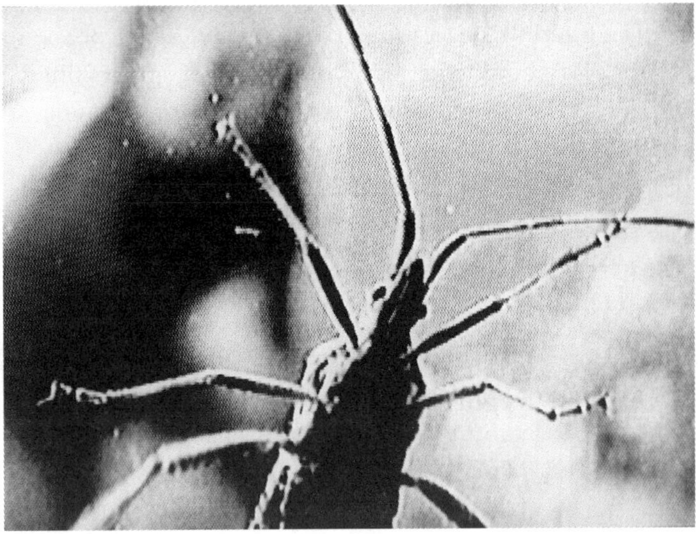

< < <

A bug crawls across Tom Sherman's camcorder lens in his video, "Wind-Up Wolf," 1999, 3 minutes and 10 seconds.

[5] It's Impossible to Tell a Story to Nature in a Language It Will Understand.

It's funny how uptight people are about telling stories about animals and plants and rocks and rivers and shooting stars. Every time people start to tell a story about nature, someone is ready to jump all over them for being anthropomorphic. That means they are guilty of attributing human qualities to creatures or things that are not human. This attribution is done most frequently through the convention of the voice-over. Human thoughts and voices are inserted into the minds and mouths of other species. Anthropomorphism seems to be particularly offensive when it is done to entertain. It doesn't matter that squirrels act like cartoon characters. It doesn't matter that all babies are cute or that all young adults fight over sexual partners and territory. But if you dare put human qualities on animals, a lot of people will think it's awful and condemn you for doing so. The defenders of anthropomorphism say people have trouble relating to nature without making it like themselves. They say if we can't make nature like ourselves, we won't be able to relate to nature at all. And that's tragic. People often feel alienated from nature. There's a problem with stories that don't feature much human interest. How can you tell an interesting story about a spider and a rock, or about a rock and another rock? It's hard to get hooked on rocks or spiders unless you give them something to say or do that you can relate to. Or you could just keep your mouth shut. If you keep your anthropomorphic stories to yourself, who's going to know? One thing for sure, nature won't care.

[6] There Are Two Conflicting Notions of Truth and Nature Always Wins.

There's scientific knowledge and then there's human instinct—our intuition, our sixth sense. For a couple of hundred years, the idea of scientific inquiry and objective truth has been gaining strength at the expense of what we know naturally. Since the close of the eighteenth century, scientists, detectives, and lawyers have been assembling evidence in logical narratives to prove beyond a doubt the truth of certain speculations. Theories are put forth and proven to be true through inquiry and the logical arrangement of data. In the information age, this has led us to believe that the planet is threatened by global warming and that violence on television leads to violence in our homes and on the streets. We would be wise to remember that truth used make more sense than it does today. We used to go to special places where magic moments revealed truths. We used to go without water to the desert, where we would see visions of the truth. Nature would provide us with the real story. It still does. There are too many humans on the planet. There is too much technology. And the humans driving the most powerful technology believe mistakenly that they control nature. I whispered these truths to myself without thinking one day while my eyes were engaged with the perfect shape of a cedar bush adorning the entrance of a Burger King restaurant near my home.

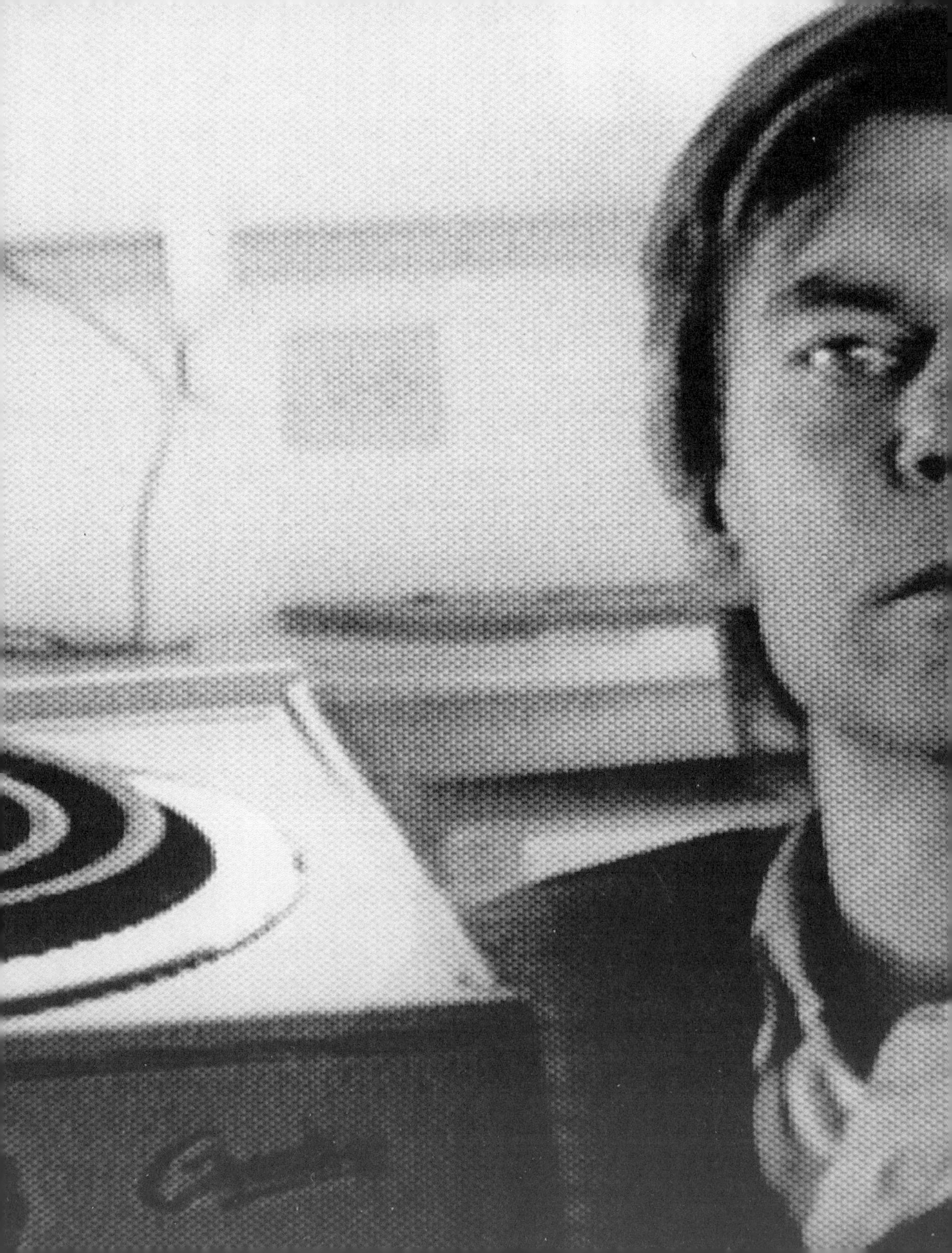

EPILOGUE > 365

TOM SHERMAN

Epilogue

Nature on earth, in the sense of untamed wilderness, is gone. There are still remote corners of the planet where you can be devoured by black flies, or a bear, but for the most part the earth has been completely overwhelmed by the human species. With world population pegged at just over six billion today, experts seem to agree that the global population will peak at around nine billion in 2060. This is supposed to be good news. Earlier estimates were pointing toward twelve billion. Whether or not the earth will be able to sustain nine billion humans remains debatable, given the self-destructive behaviour of the species. By 2060, nature will be very perverse indeed.

Back in the late sixties and early seventies, I had the good fortune to attend a couple of live appearances by R. Buckminster Fuller. Fuller was an unforgettable speaker. His monologues were often over three hours in length, profoundly nonlinear, beautifully poetic, chock-full of ideas and very funny. He did stand-up as a crackpot engineer, and when his audiences couldn't laugh anymore, he would break their hearts with his big ideas. He said it was too late to think about nature or natural ways of doing things without technology. Bucky said that everything that man did was natural. For someone like me, who was already guilty for loving science and technology during the early years of the environmental movement, this phrase was music to my ears. I knew Fuller wasn't endorsing the destruction of our ecosystem. He was acknowledging the simple fact that the future evolution of our planet, the spaceship earth, is our responsibility. The human species is the major success story of the earth's history to date, and the future of our species and all living things is in our hands.

Since Fuller stated our dilemma so clearly, the species has continued its dance with the planet's life in the balance, taking two steps forward, one step back; one step forward, two steps back. Diseases are eradicated with antibiotics, but drug-resistant bacteria emerge. Drought-based famine is countered by a global redistribution of food, only to be foiled by outbreaks of civil war.

Energy dependence on coal, oil, and natural gas is mitigated by nuclear power with disastrous consequences. Then coal, oil, and gas are embraced again at the immediate expense of the atmosphere. Skin cancers resulting from solar radiation are treated with radiation therapy. As genetically modified foods are introduced with reckless abandon, we wait for the other shoe to drop. Most would agree there is a clock ticking behind the "festivities" and this dance cannot go on forever.

We are dealing here with the highest level of drama. This incredibly dramatic "story" will be played out in wild oscillations of hope and despair, in collective emotional responses to staggering, unprecedented events. Get ready. The first cloned human may very well have interesting and important things to say when he or she grows up. If the broadest public has had trouble taking the emergence of artificial intelligence and artificial life forms seriously, when our technology speaks to us in the image of our own flesh and blood, we will surely get the message.

In the meantime, we have already bonded emotionally with a technology alien in its concrete, plastic, metallic, electronic nature. Let's face it, all this stuff in-between us — the buildings, automobiles, cameras, microphones, computers, telephony, the networks — it is so damned attractive, and necessary. Whether we're at work or home, or roaming with wireless gear while in transit or just hanging out, we are embracing technology itself as *the significant other* in our lives. Communications technology does enable us to connect with other flesh and blood people at a distance, but there is something else more physically immediate going on. The screens themselves have become the faces, the eyes we look into. And the speakers or earphones are the voices we listen to. The e-mail address books are our neighbourhoods. The daily traffic of voice, text, and image we have learned to manage is our local culture.

It is easy to sink into the sadness of this distance between people. Cell phones and networked computers are not what they're cracked up to be. In fact if we take a good look at these social technologies, they clearly double as the iconic representations of disconnection. For every potential meaningful connection, there is a much greater price to pay for all the general dislocation and down-time — an emotional cost based on disappointment, feelings of inadequacy, and mounting fatigue. When we manage our personal affairs with same technology we use to manage our clients, accounts, shipping and receiving, promotional mailing lists, and stock portfolios, should it be surprising to find that we demand efficiency and decisiveness in our conversations with loved ones, or that we feel we have to get it in writing when someone says they care?

It is hard not to dwell on the negative when one negotiates the totality of one's reality through technology. A cyborgian, posthuman future simply has the shadow of dystopia stamped all over it. It is understandable, even natural, that we find ourselves unable to appreciate the positive aspects of being inseparable from technology — high tech and low tech and everything in-between. There appears to be an inescapable myopia in our dystopian perspective.

If we want to look beyond the depression of techno-existentialism, there are steps we can take. I personally find the sweetness of irony therapeutic. There is something constructive and self-affirming about issuing clusters of grievances with a twist. With this said, you can never tell people when and how to laugh. Laugh tracks are ill-advised unless they are effectively designed to relieve and not compound the pressure of hopelessness. When the noose of hopelessness is drawn tight, even a half-laugh is beneficial.

Thankfully, hopelessness is next to impossible to render. No matter how tightly a door is closed, there is always a crack, a ray of light, a glimmer of hope. Our worlds are composed of two primary elements: memory and hope. Hopelessness is the product of excessive memory, the aspect of life that can be recounted or duplicated, ad nauseam. Hope can never be contained or terminated. It always escapes memory. We need people and technologies that seek and amplify hope. This is an important, extremely difficult mission.

If we accept this challenge, it is our responsibility to evade the constraints of memory, if only for a moment. To this end, let us make images never before seen, sounds never before heard. Let us form ideas never previously contemplated. To do so, we must become otherworldly. For the planet earth has more than enough memory to go around. The earth's ecosystem, including all the debris we've dumped in the sea, on land, and in outer space, is one hell of a mnemonic device. It simply never forgets.

With all respect to the planet earth, it's now time for something else. Go forth, benignly, go forth.

About the Author

Tom Sherman was born in Michigan in 1947. He received a B.F.A. degree from Eastern Michigan University in 1970. In 1972, he emigrated to Canada, where he settled in Toronto. He completed his post-graduate work on the job at A Space, one of the first artist-run centres in Canada. By 1974 he had begun writing to extend his concept-based work into publications, texts-as-visual objects, video, and performance. Sherman lived in Toronto until 1981, working in video, performance, and concept-based, sculptural installations.

Sherman co-founded organizations such as A Space Video (1973, Toronto) and *Fuse Magazine* (1978, Toronto). During this same period he was a regular on-air contributor to CBC national radio, appearing as a monologist on Peter Gzowski's original *This Country in the Morning*, and on *As It Happens*, hosted by Barbara Frum. He also wrote for TVOntario's ground breaking *Fast Forward* series, and produced some of the first music videos ever broadcast (1976–78) for TVOntario's *Nightmusic* and *Afterimage* programs.

In 1981 he moved to Ottawa, taking the position of Video Officer within Visual Arts Section at The Canada Council. Sherman founded the Media Arts Section of The Canada Council in 1983, becoming its first Head of Section (and establishing the Council's first grant programs for computer-integrated media). In 1988–89 Sherman worked with Simon Fraser University to develop a research institute, the Centre for Image and Sound Research (Vancouver).

Sherman was chosen to represent Canada at Venice Biennale in 1980. In 1983 The National Gallery of Canada mounted "Cultural Engineering," a ten-year retrospective of his video, installations, and writing. In 1986 he was an international commissioner for the "Art, Technology and Informatics" exhibition in the central Italian pavilion of the Venice Biennale.

Over the past three decades Sherman's work has been featured in hundreds of international exhibitions, festivals, and broadcast venues, including the Vancouver Art Gallery, the Museum of Modern Art (New York), Documenta X (Kassel), Ars Electronica (Linz), Musee d'art moderne de la ville de Paris,

LUX Cinema (London), Montevideo (Amsterdam), Tranz-Tech/Toronto International Video Biennal, and Elektra 2001 (Montreal).

Throughout the 1990's he was a frequent contributor to "Kunstradio," the weekly radio art program on Austria's national radio network. In 1997 he released "Personal Human," an audio CD of his voice/music work with composers Jean Piché and Bernhard Loibner. Sherman and Loibner have since formed a performance/recording duo called Nerve Theory, which remains active today. Besides his performance and media art, he actively publishes in periodicals, Web-zines and Internet listservs.

Sherman is currently an Associate Professor at Syracuse University, teaching video and media history and theory in its Department of Art Media Studies. He splits his time between Syracuse, New York, and his Canadian home on the South Shore of Nova Scotia.

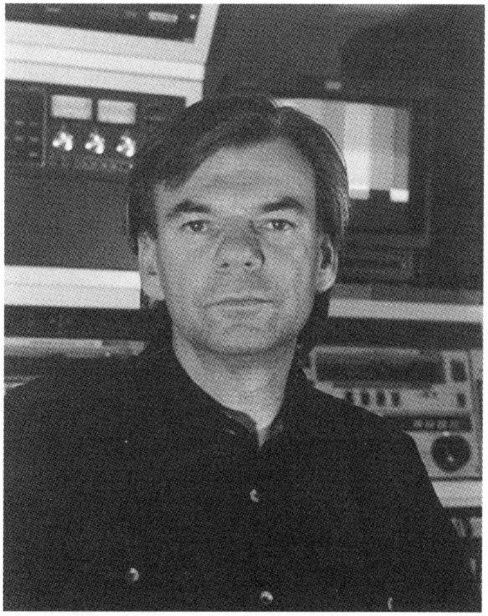

PHOTO: CYNTHIA MAYNARD

Publication Notes

Blanking________ (1996)

This is a revised version of "Blanking________" previously recorded and published on *Personal Human*, an audio CD, coproduced by *Kunstradio/ORF*, Vienna, and *Ars Electronica*, Linz, Austria, 1997.

Shades of Catatonia (1999)

This text was developed in communication and performance with Bernhard Loibner, and was premiered as a video-music-voice performance by Nerve Theory (Bernhard Loibner and Tom Sherman), September 10, 1998, at the O.K. Centrum fur Gegenwartskunst, Linz, Austria, as part of INFOWAR, the 1998 Ars Electronica Festival.

Published in *Massage*, Nomads Collective, Laura McGough (ed.) www.nomadnet.org, v. 6.0, April 2000.

TBDF (Transborder Data Flow) (1982)

Published in Tom Sherman, *Cultural Engineering*, Willard Holmes (ed.), National Gallery of Canada, Ottawa, Ontario, Winter 1983; and in Tom Sherman, *Ingénierie culturelle*, translation by Hélène Papineau, Willard Holmes (ed.), N.G.C., Ottawa, Ontario, Winter 1983; and in *Impulse* (Toronto), Vol. 10, No. 1, Summer 1982.

People Want to Know What Other People Are Thinking (1999)

This text was composed from notes taken at a symposium called "Money and Law," conducted by the New Media Institute of The Banff Centre for the Arts in Alberta, Canada, July 24–26, 1999.

Published in *ArtByte* (New York, New York), Vol. 2, No. 5, January/February 2000.

Millennial Spurn (1999)

Published in part in *Lumpen* (Chicago, Illinois), Vol. 8, No. 6, February 2000

Once Living: In a Healthy State of Paranoia (1979)

Published in *Impulse* (Toronto), Vol. 8, No. 1, October 1979; and in Tom Sherman, *Cultural Engineering*, Willard Holmes (ed.), National Gallery of Canada, Ottawa, Ontario, Winter 1983; and in Tom Sherman, *Ingénierie culturelle*, translation by Hélène Papineau, Willard Holmes (ed.), N.G.C., Ottawa, Ontario, Winter 1983.

More Dead Artists!? (1980)

Published in Tom Sherman, *Cultural Engineering*, Willard Holmes (ed.), National Gallery of Canada, Ottawa, Ontario, Winter 1983; and in Tom Sherman, *Ingénierie culturelle*, translation by Hélène Papineau, Willard Holmes (ed.), N.G.C., Ottawa, Ontario, Winter 1983; and in both English and French in *Parallelogramme Retrospective 1979*, Tanya Rosenberg (Mars) (ed.), Parallelogramme, Toronto, Ontario, December 1980.

The Value of Privacy and the Effects of Cultural Engineering (1989)

The substance of this text was originally published as "The Value of Privacy," in *Parachute* (Montreal), September 1987. This later version, revised to include more recent information, was presented at "Ethical Choices in the Age of Pervasive Technology," a conference at the University of Guelph, Ontario, October 27, 1989.

A Statement from Inside the Cultural-Industrial Compound (1981)

Published in Tom Sherman, *Cultural Engineering*, Willard Holmes (ed.), National Gallery of Canada, Ottawa, Ontario, Winter 1983; and in Tom Sherman, *Ingénierie culturelle*, translation by Hélène Papineau, Willard Holmes (ed.), N.G.C., Ottawa, Ontario, Winter 1983 and in *Impulse* (Toronto), Vol. 9, No. 2, Summer 1981.

Suggested Sway (1998)

Published in *Felix: Voyeurism*, Kathy High and Maria Venuto (ed.), The Standby Program (New York, New York), January 2000.

Television Is the History of Video (1993)

Published as "Radio Monologues," in *On the Air/Kunst im Offentlichen Datenraum*, Heidi Grundmann (ed.), TRANSIT (Innsbruck, Austria), in English and German, Winter 1994; and in part as "3 Machines in One," *Texts* (Calgary), No. 9, Winter 1993.

The Artist Attains Ham Radio Status in an Era of Total Thought Conveyance (1978)

Published in a longer version in Tom Sherman, *Cultural Engineering*, Willard Holmes (ed.), National Gallery of Canada, Ottawa, Ontario, Winter 1983; and in Tom Sherman, *Ingénierie culturelle*, translation by Hélène Papineau, Willard Holmes (ed.), N.G.C., Ottawa, Ontario, Winter 1983; and in *Centerfold* (Toronto), Vol. 2, No. 6, August 1978.

Appearance, Memory, and Influence (1986)

Published in *VIDEO re/VIEW: The (best) Source for Critical Writings on Canadian Artists' Video*, Peggy Gale and Lisa Steele (ed.), Art Metropole and V Tape (Toronto), Summer 1996; and in *Video by Artists 2*, Elke Town (ed.), Art Metropole, Toronto, Ontario, Summer 1986.

The Art-Style Computer-Processing System (1974)

Published in Tom Sherman, *Cultural Engineering*, Willard Holmes (ed.), National Gallery of Canada, Ottawa, Ontario, Winter 1983; and in Tom Sherman, *Ingénierie culturelle*, translation by Hélène Papineau, Willard Holmes (ed.), N.G.C., Ottawa, Ontario, Winter 1983; and in *Videation* (Richmond, Virginia), published by Bob Martin at Virginia Commonwealth University, Spring 1977; and in *Journal for the Communication of Advanced Television Studies* (London, England), Vol. 2, No. 2, Fall 1974.

Machines R Us (1994)

Published in *FLESHFACTOR: informationsmaschine mensch*, Gerfried Stocker and Christine Schopf (ed.), Ars Electronica 97 (Linz, Austria), SpringerWienNewYork, in English and German, September 1997; and as "As Máquinas Somos Nós ...," in *A Arte No Século XXI: a Humanização das Tecnologias*, Diana Domingues (ed.), Editora UNESP Fundação (São Paulo, Brazil), Fall 1997.

Memory and Hope (1996–2006)

This is a text adaptation of a presentation given under the title "Emerging Technologies: Options for the Near Future," at the New York State Media Festival, held at Syracuse University, Syracuse, New York, April 1996.

Published in *The Squealer* (Buffalo, N.Y.), March 1997; and in *MUTE: Digital Art Critique* (London, U.K.), Issue 6, Autumn 1996.

Nature Is Perverse Sometimes (Version 6.1) (1994)

Published in *Memesis: the Future of Evolution*, Gerfried Stocker and Christine Schopf (ed.), Ars Electronica 96 (Linz, Austria), SpringerWienNewYork, in English and German, October 1996.

Jack's Eye Was a Camera (1991)

Published in *Arts Magazine* (New York), Vol. 65, No. 6, February 1991.

From a Reservoir of Predictions (1990)

Published as a monograph initiating the Art Metropole series "Activating the Archives," Art Metropole, Toronto, January 1990.

The Face of the Machine (2000)

Published in *Austria Kultur*, Austrian Cultural Institute (New York, New York), Vol. 10, No. 3, May 2000.

Vision Beneath Light (1997–98)

The ideas in this text were delivered in their initial form at Kunstradio's Recycling the Future IV symposium in Vienna, December 5, 1997, on the occasion of Kunstradio's tenth anniversary. Kunstradio is a weekly radio program composed of radio art broadcast by the ORF, the Austrian Broadcasting Corporation.

The addendum to this text was published in a modified form as "Trouble Signs," in *ArtByte* (New York, New York), December/January 1998–99.

The Finished Work of Art Is a Thing of the Past (1995)

Published in *C* (Toronto), No. 45, Spring 1995; and in *Actes/Proceedings, ISEA'95* (Montreal), September 1995; and as "L'oeuvre d'art <<achevée>> est un concept du passé" in *Où va l'histoire de l'art contemporain?, l'image et l'Ecole nationale supérieure des Beaux-Arts* (Paris), 1997.

Museums of Tomorrow (1987)

Published in *Parachute* (Montreal), in English and French, March 1987.

Exclusive Memory (Excerpt #8) (1987)

Published in *By the Skin of Their Tongues: Artist Video Scripts*, Nelson Henricks and Steve Reinke (ed.), YYZ Books (Toronto), November 1997.

The Monologue... (1997)

Published in *By the Skin of Their Tongues: Artist Video Scripts*, Nelson Henricks and Steve Reinke (ed.), YYZ Books (Toronto), November 1997.

Equidistant Relationships (1989)

An excerpt from this text was published in *Interior Presence: Projecting Situations* (exhibition document), Amy Gogarty (ed.), Nickle Arts Museum, The University of Calgary, Alberta, Winter 1990; and a longer version in *Exclusive Memory* (exhibition document), Gary Dufour (ed.), Vancouver Art Gallery, Vancouver, British Columbia, April 1989.